SHEEPKEEPING ON A SMALL SCALE

A complete guide to the buying and maintenance of a small flock for home wool production.

By the same author

CARE AND SHOWING OF THE HEAVY HORSE
CORNWALL
DRY STONE WALL HANDBOOK
HARNESS HORSE
HEAVY HORSE
HEAVY HORSE AT WORK
HEDGE LAYING AND FENCING
THE HILL SHEPHERD
NORTH COUNTRY FARM ANIMALS
THE SHEEP DOG: ITS WORK AND TRAINING
(with Tim Longton)
SHEPHERDS CROOKS AND WALKING STICKS
(with David Grant)
VICTORIAN AND EDWARDIAN FARMING

SHEEPKEEPING ON A SMALL SCALE

Sound Advice on Home Wool Production

by

Edward Hart

THORSONS PUBLISHERS LIMITED
Wellingborough, Northamptonshire

First published 1979
Second Impression 1981
Third Impression 1982
Fourth Impression 1984

ISBN 0 7225 0517 5 (hardback)
ISBN 0 7225 0516 7 (paperback)

Printed and bound in Great Britain

CONTENTS

FOREWORD

If you are considering keeping a few sheep, start with a sense of humour at your own shortcomings because these woolly, cuddly creatures have minds of their own. In fact, they can drive you up the wall while they are going over one five fields away. It has been said that sheep can bring out the best and the worst in man in half an hour, but a healthy bunch of ewes with off-spring at foot is surely a sight to behold; and profitable too.

Edward Hart has given an accurate and practical account of the seasonal and routine jobs which are a must whether you handle a small or large number of sheep. To me there is another aspect of keeping sheep equally as important as the routine jobs. Over forty years ago an old shepherd who played a large part in advancing the Swaledale sheep breed said to me in a broad Dales dialect: 'Tha knows lad thar's plenty o' folk lukes at sheep an niver sees 'em.' I am sure that all good stockmanship begins with that observation. The ability to see at a glance either individual sheep or the whole flock, their conformation and general condition, and relate all that to the prevailing weather conditions and to the amount of grazing available. The person who can detect a decline in the animal's condition before any damage is done is the one who 'looks and sees'.

For many years Edward Hart has been writing articles and books on all aspects of sheep, sheep management and crafts associated with shepherding, and his ability to assess the balance between theory and practice is a guarantee that this book makes factual and interesting reading.

RAY DENT
Shepherd of the Year 1976

1
BASIC SHEPHERDING

Welcome to the sheep world! It is a grand brotherhood, a link with Biblical and medieval history, with most countries of the world, and with a race of men and women who lead interesting and enviably uncrowded lives. If you have but two sheep, that is common ground enough to discuss things with the shepherd of a thousand ewes.

More practically, sheep provide wool, skins, meat and breeding stock. They are self-perpetuating, so once the initial number of ewes has been bought, there is little need for future female purchases. Sheep will survive on grass and hay, though some corn is advantageous at certain times, and they fit well into any self-sufficiency scheme. Unchanging in a changing world, they are a source of continuing satisfaction, and to go and look at his sheep has been the relaxation of many a farmer harassed by the other complex departments of his trade.

Though the main emphasis of this book is on wool production, there may be occasions when slaughtering is necessary. Home slaughtering is illegal without a license. Fewer licensed slaughter houses are now available (there used to be one in every village) and owners may have to send their sheep to an abbatoir. Check that all the offal, especially the liver, is returned.

The Basics

We will start with basic principles. The sheep is a quadruped whose legs enable her to climb the highest hills, and often to jump any sub-standard fence! She is a ruminant whose stomach converts vegetable matter, especially grass, into meat and wool. If man had invented the ruminant's stomach, it would be ranked as the Eighth Wonder of the World. Yet we tend to take for granted that remarkable apparatus and its ability to convert grass, which is of little direct use to man, into warm fibres to protect his body, and meat to feed him.

Like other livestock, the sheep has disadvantages. It is gentle and cannot protect itself against predators, the most serious of which in Britain are stray dogs or otherwise well-behaved pets. The sheep is easily alarmed, rushes away, and, even if not physically attacked, is so distraught that she may never rise again, and if she has lambs inside her they may be lost.

This lack of the will to live must be remembered at all times. So must the sheep's susceptibility to a host of parasites and diseases. However, most of those that threatened our great-grandfathers' flocks are now controlled by science and routine injections. This important aspect is dealt with in Chapter 9, but basically a couple of injections and a few worm drenches annually safeguard those aspects, and the novice need not be put off by the daunting list of sheep diseases.

Sheep are shorn once a year. There are local exceptions in the south-west and with a very few breeds, but generally speaking the fleece is removed when the hot sun has helped the 'rise' in the wool discussed in Chapter 10.

Most sheep lamb in the spring. Again, there are exceptions, and the small keeper may choose a breed like the Dorset Horn which will lamb more than once a year. Lambing is a natural process, and not one to alarm the uninitiated if the guidance offered here is accepted.

Their relatively small size makes sheep suitable for women and children. The shepherdess of legend has her modern counterparts in the many women who excel at handling these gentle creatures.

Before embarking on purchases, the would-be sheep keeper must become familiar with fundamentals. One sheep could conceivably be kept alone, but like most other stock they tend to do better in company. Unlike cattle, it is not easy to keep them healthy on the same area of ground year after year, without resting it from them. Modern anthelmintics – a horrible word meaning worm drenches – help control the parasites which cause 'sheep sickness', or that state of the pasture that occurs when sheep never do very well although they have masses to eat.

The sheep is a selective grazer. She prefers fine to rough grasses, and eats them so continuously that their growth is affected (See Chapter 8). This problem can be overcome either by grazing with cattle, which are much less particular, or by regular ploughing and reseeding. But one great virtue of the sheep is its ability to thrive on steep banks quite unsuitable for ploughing.

Feet and Teeth

The most important parts of a sheep are its feet and its mouth. Never buy lame sheep. The mouth must be sound enough for the purpose in hand, which usually is to cut off grass and ingest it. The teeth we are concerned with are the incisor or broad teeth of the lower jaw. These meet a hard pad on the top jaw. If they do not, the animal cannot eat properly. One of the first things a shepherd must learn is the difference between a sound and an unsound mouth.

The molars concern us less. In Iceland the shepherds note them closely, probably on account of damage from the larva-strewn pastures, but in Britain the lower incisors are most important, not least because they indicate a sheep's age.

When a lamb is born it has eight front milk teeth in the lower jaw. The age at which these are displaced by permanent or 'broad' incisors varies with breed and nutrition, but generally speaking the two middle broad teeth have erupted at eighteen months. At about one year nine months the sheep has four broad teeth on the lower jaw, and at two years three months the sheep has become a 'six-tooth', a descriptive term used in some areas to denote her age. The change to eight teeth or a 'full mouth' occurs between three and four years of age.

Occasionally sheep do lose an odd tooth through accident, such as stone damage while grazing, but this is not a serious way. After the full-mouth stage, more can go wrong. The incisors may grow long and project forward from the pad of the upper jaw. The owners of such teeth are then inefficient grazers, and are better left to someone experienced with this class of stock.

Names

Names for the various classes, ages and sexes of sheep seem unbelievably complicated at first. They also vary in different areas. We will start with basic terms of wide acceptance. The adult uncastrated male is a *ram* or tup. The act of mating is 'tupping'. The adult female is a *ewe*. The newly-born youngster until the end of the year in which it is born is a *lamb*, but these terms are not always used on their own.

A male lamb is a *tup lamb* or *ram lamb*, unless it is castrated, when it becomes a *wether lamb.* A female lamb is a *ewe lamb* or *gimmer lamb.* After 31 December of their first year, these lambs become *hoggs*

Market scene in the Borders. These little Cheviot lambs are still bred in their thousands, despite much of their better grazing being taken over by forestry interests.

in the north and *tegs* in the south, retaining their appropriate sex distinctions; *gimmer hogg, wether hogg, tup hogg.*

Following the hogg or hogget stage, the sheep becomes a *shearling*. This is a literal term; it is so called after it is shorn of its wool for the first time, which is about July of the year following its birth, i.e., at some sixteen months old.

The shearling age lasts until the sheep is shorn again, at just under two and a half years old, when it becomes a two-shear. The following year it is a three-shear, then a four-shear. In each case add a year to find the age in years. The same rule applies in districts where sheep are described as 'two-, three- or four-crop'. They have had that number of lamb crops, and as the first was at two years old, add a year to the number of crops to find the age in years.

A shearling ewe is termed a *gimmer* or *gimmer shearling* in the north, a *theave* in the south. Don't worry if you can't remember all the terms; lots of farmers know only those used in their own area and they buy and sell sheep well enough.

A pleasant term which will seldom concern us is the Cumbrian name *twinter* for a sheep in its second winter. Herdwicks on the highest Lakeland fells do not produce their first lambs until three years old, because of slow growth on exposed, thin soil.

Wether Sheep

More important to the self-sufficient shepherd is the term *wether*. We have seen that a wether lamb is a castrated male lamb. *Wether* on its own refers to an older castrated sheep, possibly coupled to age definition, e.g., four-shear wether.

There is considerable virtue in the wether sheep for novice shepherds mainly interested in wool. Last century, and to a lesser extent in this, wether flocks formed an integral part of hill sheep farms. The wethers were kept till five or more years old, solely for their fleeces until they were eventually sold for tasty mutton. Being of good age and born on the place, they were experienced in the ways of the weather and the farm, and acted as flock leaders. They marched the unfenced boundaries and kept neighbours' sheep at bay. Most importantly, they led the breeding ewes to safety in a storm. They declined in numbers as wool prices dropped and demand for fat mutton ended, being replaced by ewes and lambs.

On the more exposed Scottish hills there have been *wether hirsels* (see Glossary) until quite recently, and for the same reasons that commend the wether to the newcomer. Wethers are mature sheep, and so have pased the danger stage for many lamb-killing diseases. They resist intestinal worms better and require less shepherding throughout the year. Being non-breeders, they have no breeding problems. At no time of the year do they call for the better grazing and improved shelter associated with a flock of lambing ewes.

If a few wethers are decided on, special care in choice of breed is necessary. The sheep is not going to provide an annual lamb crop for sale or consumption. He will become fat, but you may have become so fond of him that you do not wish him to go to the butcher! So he may as well earn his keep by providing a really heavy fleece.

The widest selection of wether lambs, or indeed of any lambs, comes in autumn. This is the best time for the purchase to be made. Wether sheep of two, three or even four years old are sold at a few sales, mainly in the remoter hill areas, but are usually ready for the butcher. This does not make them unfit to keep longer, but it does mean that you pay for meat laid on the carcass that you don't really want. Wethers for the butchers are not cheap, as people have realized the virtue of their meat. On the Isles of Lewis and Harris off Scotland's west coast, no grain is grown and the male lambs are castrated and kept until they are fat from grass and hill herbage. The result is a four-year-old wether whose meat is of delicious flavour. New Zealanders prefer wether mutton to tender, tasteless lamb, and Her Majesty Queen Elizabeth II was served with Herdwick wether mutton at her Coronation feast.

Sheep Sales

At most sales the main consignments are in pens of a score or a hundred lambs. Yet there are always the few stragglers which do not match the rest. These are sold on their own, as are the shepherd's 'pack' sheep in Scotland. These are the employed man's own property, usually of good quality but small in number.

Having seen a likely batch, the buyer's first business is to check that the sheep have indeed been castrated, that the job has been done properly and that no infection of the 'cod' – the scrotal sac minus testicles – has occurred. If correctly carried out, castration results either in a clean skin or an empty, skinny scrotal sac according to the method used.

If stock are sold as wether lambs or 'clean cut', that is a warranty that castration has been properly carried out. It is still up to the buyer to check: 'Let the buyer beware!' or *caveat emptor* in the old Latin tag. Far better make sure at the time, for putting right such matters afterwards is a time-consuming and aggravating business disliked by all concerned.

Self-sufficers may well produce more lambs than they need, in which case they may be sold privately or at one of the auctions which take place throughout the autumn.

The wether lamb will need no corn after its first year. Two multiple injections a year, and the same number of worm drenches, will protect it from most diseases and internal parasites. It will thrive well on second-rate pasture, and will produce a heavier fleece than a ewe of the same breed as it is not concerned with the strain of bearing and suckling a lamb.

Simplicity of management should not be confused with neglect or under-feeding. A reasonable level of nutrition throughout the year should be the aim with the wether. An excess or deficiency shows itself in the wool, which has weak places to correspond with stress or malnutrition. In his *Complete Grazier*, 1852, Youatt wrote: 'It is essential for the evenness and strength of the staple, that the feeding of the animal should be uniform, and without any sudden interruption or transition.'

Much of the world's fine wool comes from the Merino breed, and quite a proportion of that from wethers, left largely to their own devices on the stations of Australia, or climbing to the New Zealand snow line. Unfashionable in Britain for over half a century, this class of sheep may again come to the fore through the efforts of self-sufficiency enthusiasts.

2
BUYING SHEEP

The usual advice to amateurs buying livestock is to approach a reputable dealer. This counsel is not always easy to follow through. If you are a stranger in a strange land, you will know no dealers, reputable or otherwise. In long-settled country areas there is the probability that the advisor is not wholly disinterested; his nomination may be his wife's cousin who is prepared to share his commission! Country dealers are like the rest of the human race; some will help a lame dog over a stile while others would take the buttons off the shirt on your back.

The best solution, if you have no reliable contacts at all in the sheep world, is to join the National Sheep Association. For a few guineas a year the same service will be open to you as to the thousand-ewe man. Wherever you live in Britain, a link will be suggested to guide you through the first uncertain months. Far better pay a little more for advice from a distance than to lose the lot through having no advice at all.

If a neighbouring farmer proves helpful from the start, he may be persuaded to part with a few sheep from a pen that he is buying or selling. He may take you to the market when he bids, and if he charges a pound a head or more for what seems a few minutes' work, don't quibble. He is selling his years' experience of sheep, sheep farms and men.

Most sheep change hands at markets, and usually in bigger pens than the newcomer requires. Sometimes a ewe and her two lambs are sold as a separate lot. Beware here: they may belong to a smallholder short of keep or money, but they may equally be experienced fence breakers that no ordinary barrier will keep at home. To buy the latter is to experience hours of frustration.

Time of Year

Autumn is the easiest time of the year to buy sheep. There are sales every week, perhaps every day, within easy radius. The majority of British lambs are born in the spring, and most of those not fattened during the summer are sold in autumn at store sales (see Glossary). Also sold are breeding sheep of all ages, the majority of which are draft ewes from the hills. These are middle-aged sheep with one, and probably considerably more, years of breeding in them. Draft ewe sales are held in all upland areas. The other main sales are of rams, and ewe lambs or shearlings for breeding.

After the lamb sales, the ewe sales. Here Scottish Blackface ewes are off to a new home.

Age

The sheep's maximum value is at two to three years old, i.e., as a two-shear. Few change hands at this stage, and for beginners younger sheep are not ideal. Well-grown ewe lambs of most lowland breeds may be mated or tupped in their first year, i.e., when they are about six to ten months old. They have never lambed before, and neither has the novice shepherd, so why make life difficult? Shearlings are even worse. These are put to the ram at about eighteen months old, to lamb down when they are two years old. Shearlings that produced lambs in their first year have gained a little sense, but to spot them requires expert knowledge. For some strange reason, maiden shearlings are more stupid than yearlings at their first birth. They are more likely to have twins, which they may look at and leave. Far better start with older, but not too old, ewes. Those of four to six years old are recommended.

Ewes of this age are only just past the prime of life. They are experienced dams; they know far more about the facts of sheep life than the newcomer. They are sold from the high moors or open hills to the rather more sheltered dales or lowlands, and any enclosure round a house is likely to provide more amenable conditions than the exposed heights, especially in winter.

Sheep have a reputation for living long off particular hills, whereas from others of different soil type and wetness they survive less well. This is the sort of knowledge paid for through a friendly buyer or, more expensively, through burying carcasses. Generally speaking, the drier type of hill, especially in a limestone area, produces sounder stock than a wet one.

Don't buy really old sheep of more than six years. Again, they are for the expert, who can nurse them through the toils of winter and lambing. With novice handling, they are only too likely to give up the struggle, and refuse to battle for their lives or lambs. 'Old horses and old sheep are the road to bankruptcy', said a between-the-Wars farmer.

After sheep become four years old, they are termed full-mouthed. Precise age definition then becomes difficult, but this matters less than the condition of the teeth. When these become long and overgrown so that they do not meet the upper pad properly, the mechanism of grazing breaks down, and the animal fails to thrive.

In some hill flocks the year of birth is stamped on the horn. A sheep born in 1979 has the figure 9 as a brand. Watch out here; occasionally the practice is to use the year in which the branding is done, and this is at the hogg (see Glossary) stage or when the sheep is one year old. Thus the figure shown indicates the age in years plus one. It is uncharitable to suggest that the latter practice amounts to deliberate deception, but let the buyer beware.

Ewes

Ewes are sold as warranted or unwarranted. In the former case they are guaranteed to be of the age stated, and to be sound 'above and below'. This means correct in teeth and udder. Teeth have been discussed. A healthy udder is soft and pliable, with both sides even and free from lumps. If one quarter is much larger than the other, the sheep is suspect, and very large 'balloon' teats should be avoided. These enlarge so much when filled with milk that the newly-born lamb is unable to suckle, and has to be helped on each occasion. Unfortunately, it is impossible to guarantee a healthy udder the following spring. the tissue may be perfectly sound in autumn, but become infected during the next six months.

Ewes With Lambs

If ewes are bought in spring with lambs 'at foot', the udder may be more easily checked. White milk should be drawn easily from each teat, both should appear to be functioning evenly, and the condition of the lambs should denote that all is well. Be quite certain of the terms of auction when ewes and lambs are sold together. At some marts the price refers to the ewe and her lambs together, while at others the figure called by the auctioneer is for each animal. In that case a ewe with twins selling at £15 would total £45.

All classes of livestock tend to be dearer in spring, when abundant grass is available, than in autumn with an expensive winter ahead. We have seen that autumn is the best and easiest time to buy sheep, but if you have grass available in spring, there is no point in waiting. Both grass and time would be wasted. Ewes with lambs at foot, or 'couples', are fairly straightforward. They should be avoided if the lambs are under a week or ten days old, but after that time it should be comparatively plain sailing.

In autumn, if you are lucky, the lambs may be sold for the purchase price of the lot, in which case the ewe remains as a bonus. The lambs may, of course, be retained either as future breeders, if female, or as wethers if male. Or the lot may be sold and new sheep bought in, a fresh choice of breed having been made in the light of experience.

Rams

Ram buying is a yearly, or at least biennial, operation. The small flock owner is unlikely to be able to retain his own males, as they would be too closely related to the existing stock. He therefore goes to a tup sale. As a means of mixing business with pleasure, these events are unbeatable. Sheep farmers, and especially hill farmers, enjoy their infrequent sojourns among their fellows, and never fail to take the opportunity to catch up with gossip and good living. The novice buyer is even more likely than the professionals to get carried away by the excitement of events, and to bid for something he can't really afford, and could do without if he considered the matter seriously at home.

Although there is no cure for tup sale fever, a few basic rules apply to buying the flock sire. He is indeed half the flock, for his characteristics will make up half of those of the next year's lamb crop. There is no need to pay a high price of several hundreds of pounds, for there is no guarantee that an expensive animal will sire better stock than a much cheaper one.

The ram should be pure bred, and of good breed type. Lowland rams are comparatively easy to procure. If a Down ram is needed to run with some draft or halfbred ewes, there are hundreds to choose from, at little above commercial prices. Length, depth and fleshing qualities are needed in the Down ram, and such is the skill of breeders that these points are there in abundance.

The Fleece

Where sheep are being bred pure, and replacement females retained, be very careful about the ram's fleece. Wool is highly heritable; as the parents are, so will the children probably be. One experienced hill farmer put it this way: 'There are lots of things to go wrong in breeding, apart from wool. If you buy a ram with the correct fleece, that is one less factor to worry about.'

The fleece must be correct for breed type. If of a white breed, it must not have an intermingling of black or red fibres. Small black spots in other breeds, especially at the tail head, down the hind legs or behind the head, are less important. It is easy enough to describe the perfect tup according to his breed; the buyer's skill is in selecting a serviceable one in his price range, and one that has less important faults.

An important principle is to buy a run-of-the-mill animal from a sound and noted flock, rather than a flashy individual of no known breeding. Ram breeding is a specialized business, with breeders turning out a score or more of rams for sale. Some of them disappoint the vendor in price, for no apparent reason.

Such men are concerned with next year and the year after. It does not pay them to pull a fast one over a new customer. Rather do they try to supply at a price the buyer can afford, hoping for more sales in years to come on the basis of successful matings.

Shearling Rams

Shearling rams are the most expensive. They look glorious. They are the pick of perhaps several hundred ram lambs, yet in most cases are unproven. An older ram of three- or four-shear is better for the beginner. They are especially suitable on young ewes, though on older females that know their job, a younger ram may be tried.

'Movement of Animals' Record

A Movement of Animals record is compulsory. Printed sheets are available, with columns for date of sale or purchase, particulars of the animals, and destination when sold. These books or sheets are inspected periodically by the local police. They become vital in case of an outbreak of foot and mouth or some other infectious disease. Keep them right up to date. This takes very little time, but will be appreciated by your local officer.

3

FENCING, HOUSING AND YARDS

'Good fences make good neighbours' is an old and proven farming saying. Don't get the sheep *before* the fences or you start on the wrong foot. The main principle of fencing is to keep stock on the right side. Once they develop the breaking habit, control is much more difficult.

The type of fence depends to some extent on breed, and on whether cattle and ponies are also being fenced against. Lowland sheep breeds are usually easier to control than hill breeds, and some of the halfbred ewes out of hill stock are notorious fence breakers.

If you are in stone wall country and the walls are good, nothing else is needed. For moderate walls, a small net or strand or two of wire may suffice; farmer neighbours will advise here.

A really well-laid hedge is sheep-proof, but such hedges are very few and far between. Usually a light net must be run along the front, tied every other yard or so with baler twine begged from a farm or saved from the hay crop.

Wire Fences

There are three main types of wire net. One is the lighter sheep net of fine wire twisted on the joins, and used for movable folds. It suffices for that purpose and as a hedge strengthener, but not as a permanent fence. For that purpose the sort classed as pig wire is best. It has larger squares at the top, small ones at the bottom. Don't make the mistake of setting it upside down! The smaller bottom squares prevent lambs escaping.

The third type, known in the trade as sheep fencing, is useless. Its squares are far too big, and any self-respecting Herdwick or Welsh could escape in a matter of minutes.

Post-and-rails fencing is of course excellent, provided the bottom rails are close enough or a light net is set alongside when lambs are young. But it is too expensive for most.

The other type of wire fence is the strained fence composed of a number of strands of plain wire. This is most suitable over long distances and fairly smooth surfaces. It needs some skill and experience to set it right, otherwise it is worse than useless. Really heavy anchor posts are needed at the corners, set three feet into the ground and abutted by long leaners, strainers or 'godfathers'. If a short leaner is applied, the main post is simply levered out of the ground as the wire is tightened. We will leave this type of fence for larger-scale operators. It is cheaper per yard than the wire net, given a long run; for short runs the extra costs of corner posts cut out any cost advantage.

The wire mesh fence is expensive, though when New Zealand was being settled a roll cost as much as a sheep, so it is still a viable proposition here. The posts, stakes and plain wire for the top cost about as much as the mesh, and staples are by no means a negligible item.

Tools

Tools are simple and few. Fork, scythe and slasher for clearing the ground, spade for digging holes, crowbar – to insert before the stakes are driven home with a sledge hammer – saw, pliers, hand hammer, nails, staples and some form of wire tightener suffice in most cases.

Setting the Fence Line

First task is to determine the fence line. If it is an internal fence, there is often a case for taking a new line away from the zig-zaggy stumps that may have formed the now worn-out demarcation. If erecting a boundary fence, the greatest care must be taken to ensure that the line is exactly right. A well-set fence lasts at least ten years, and unnecessary moves are profitless.

Set the line dead straight as far as possible. Changes in direction are a source of weakness, as the wire tends to pull over the uprights. Clear away any bushes, long undergrowth or overhanging branches. Stumps are sometimes useful for nailing down onto, but in general the fence path should be as smooth as possible. Though it can be steeply sloping, as straight a line as possible is more important than ever.

End and Corner Posts

Then set the end or corner posts. A depth of two and a half feet is recommended, and posts at least six inches square are needed. Big, rough hardwood branches are quite suitable if reasonably straight. They do not need to be sawn smooth. A nick is made into the face of the post, high up and in the direction of the fence. Into this a leaner or godfather is inserted, the lower end being dug into the ground and faced onto a big stone or clog of wood.

This is done at either end and at unavoidable changes in direction points. The net is then fastened very securely at one end and run out to the other. Two or three stakes are lightly driven in *en route* to lean the net on and prevent it flopping about during tightening.

Straining the Wire

Then stretch the net from the free end. This may be done by threading a metal pipe through the end meshes, and tightening the pipe. There are various methods. The chain and lever gadget is effective, or a metal crowbar with the curved end used to gain purchase. Whatever is used, something rigid to tighten from is essential. It may be a nearby tree or a tractor, or a post may have to be set specially for the purpose, and the top joined by leaner to the foot of the fence post.

Setting up the straining-posts and straining the wire is literally half the battle. Once this is done properly, the rest is comparatively easy. Stakes are needed every ten feet or so, using a piece of wood of the exact length. A fence with unevenly-spaced stakes looks amateurish, and is.

Each strand of wire on the straining posts must be securely fixed with two or three staples driven home. On the intermediates, far fewer are needed. If the mesh has eight horizontal wires, they may be stapled on Nos. 1, 3, 5, 7 and 8 on one stake, at Nos. 1, 2, 4, 6 and 8 on the next. Nor are the staples driven right home on these intermediates.

The fence foot is a vital part. In slack places, the wire is dropped as far as possible, but it may still be necessary to make up ground with flat stones, lengths of wood or a piece of plain wire between stakes. Above, one, or possibly two, lengths of plain wire are set three inches above the net. These also must be tight. Avoid barb if you possibly can.

Driving In and Finishing off

When driving in the intermediate stakes, a hole is first made with a pointed crowbar. Then the stake is tapped lightly in, ensuring that it is upright, till it grips. An assistant keeps it upright by pressing at it with a square-ended board, keeping hands well clear of the driving mallet.

Finally, saw off the tops of the stakes to an even height, using a sharp saw so that the stakes are not disturbed. This touch puts a finish on the fence. Tidy all cleared brushwood, pick up all tools and spare wire, especially any short lengths, which are highly dangerous to stock.

It is possible to buy wire tighteners for individual wires, the idea being that they are tightened yearly. In practice they are often left until too rusty to move. But they are useful if not neglected.

Changes of Height and Preserving

For changes of height an extra stake is often needed. It is driven into a low place and the fence stapled down to it, or on a bulge it lifts the net to the required height. Stakes are best if creosoted under pressure, and this may be done at home by boiling them in a barrel of creosote, and allowing them to cool before repeating the process. On a small scale the economics are doubtful.

A 'roven' stake lasts longer than a sawn one. The former is produced by splitting a length of timber with wedges into appropriate thicknesses, then sharpening the ends. I have seen oak stakes made in this way that will last more than a lifetime on a dry bank. Ash rots at ground level. Sycamore, elder or hazel are quite useless for stakes. Larch is good, and creosoted pine thinnings last for some time.

Housing

The sheep is endowed with a fleece that enables it to withstand the worst British winter without a roof over its head, provided it is adequately fed. In Iceland that does not apply; conditions are so severe that housing is universal. Interest has grown in the practice here, and the Icelanders' experience has been put to good use. The main reasons for housing in Britain are for the sake of the land and the shepherd, not the sheep.

If you wish to stock really heavily with sheep during the grazing season, it heips considerably if the sheep are moved elsewhere for the winter. The pasture is not then 'poached' or damaged by hoofs in wet weather, and also has a break from worm eggs. A shepherd can look after more sheep, or the same numbers in less time, if he has them under cover with everything to hand.

Housing of sheep is possible, provided adequate ventilation is given. The basic principle is to ensure that the temperature inside is not much different from that outside. If the house becomes too warm respiratory troubles follow.

A compromise is a slatted outdoor area with high, sheltering sides. Becoming popular in Ireland, such a system is more practical away from the areas of heaviest rainfall.

To house sheep successfully, a few basic principles must be followed. For medium-sized sheep, allow at least eleven square feet per ewe. Remember that a batch of ewes may appear to have plenty of room in autumn, but by the time spring arrives and they are heavy in-lamb, they may appear cramped. This applies especially to trough space. It is the controlling factor, even more so than square footage, and for a medium-sized ewe like a Clun, eighteen inches per sheep should be allowed.

The number of sheep in a batch should not exceed forty. With that quantity of sheep in a pen, all remain quiet, but with larger lots there is more danger of ewes becoming unsettled and dashing about if anything strange happens. More lambing problems occur with large batches.

Draft hill ewes that have spent their first four or five years on open ground do not always take kindly to confinement. I believe this question of habit is more important than breed; on one Ayrshire farm, housing was not completely successful until all sheep had spent their formative years on the system. They were the same breed and strain as their predecessors, the difference being that they had become accustomed to housing every winter.

Flooring

It is possible to house sheep on slatted floors with no bedding, but the price of timber is such that the benefits may not be worthwhile. If a layer of hard core is used, topped with straw and shavings for bedding, little dung will be accumulated and the annual clean-out becomes a quick operation.

Troughs

Troughs are important, whether inside or outside. The same trough will serve for hay and concentrates if properly designed. Those conventional hay racks in which the sheep thrust their faces upwards between iron rods are an abomination, because hay seeds fall into the fleeces around the neck and shoulders, and are impossible to get rid of at home, and expensive to counter by the trade. Furthermore they are wasteful; a sheep pulls a mouthful, stands back, and proceeds to drop the parts not actually in her mouth onto the floor.

An effective and simple trough has been designed by farms manager Tom Fulton at the Institute of Animal Physiology, Babraham, Cambridge. Made in six feet sections, the trough is of sheet metal and angle iron. It stands on eight inch legs, the trough sides being nine inches high. Above the side is a twelve inch gap topped by a rail, with another seven inch gap to the top rail.

Most important, the trough base has rounded corners where it joins the sides, following Mr Fulton's observations of sheep feeding from a half-round glazed trough: 'They all stood up to the trough, and as they ate, the surface curve caused the hay to be pushed back into the

The author (left) discusses a very practical sheep trough with Animal Research Council farm manager Tom Fulton. The rounded corners along the trough floor help the sheep to eat every scrap.

centre', he told me. His metal troughs have the triple function of feeder, pen divider and inspection platforms. Being able to inspect housed sheep without actually going in among them is a tremendous advantage and a great help to that basis of all stockmanship, observation.

Ventilation and Temperature

Housed sheep must not be in a draught, yet they must have fresh air circulating above them. This is achieved by solid walls several feet high, topped by open or mesh sides. Ridge ventilation is also important; in Iceland automatic ventilators may be installed, but for British conditions a gap along the ridge usually suffices. Mr Henry Fell, who houses large numbers of sheep on a Lincolnshire arable farm, insists that the indoor temperature should be as near the outdoor temperature as possible. In most systems of housing, ewes may lamb inside, but are moved out fairly soon.

Sheep Yards

A lot of time and effort has been channelled into designing sheep yards with drafting races, working races, foot baths and circular forcing pens. For the person with a score of sheep, such refinements are unnecessary. A series of small pens, preferably opening into each other as well as into an enclosed yard, should suffice. Vendors of portable sheep handling equipment do not always cater well for the shepherd who wishes to pop a certain animal into a pen for a few minutes whilst he or she decides what to do with it. And it is well worth spending a few extra minutes searching memory or records before deciding which ram is to have the addition of a particular ewe to his harem.

A small yard opening from the adjacent paddocks is a great asset. If sheep grazing is distant from the homestead, it may be worthwhile investing in or making portable yards, which are basically interlocking hurdles. Once in the yard, the sheep are easily driven into smaller pens for working on, or examination.

Pen sides should be clean, and strong. Wood, brick or stone are all suitable, some of the best pens being of dry stone wall fashioned by craftsmen. A height of 3½ feet (106cm) is usually enough. Even ten sheep in a pen exert a surprising amount of pressure, and it is vital that posts be heavy enough and well enough set to withstand them. A shaky fence is dangerous, and to have a mob of sheep escape halfway through dipping or shearing does not endear one to the ovine race, despite the fault lying with the human element.

Yard Floors

Yards floors should be concrete or hardcore. At Hawick, Roxburghshire, claimed as the country's oldest market, there are pens with bottoms of smooth, rounded stones, which are washed clean every time it rains. Concrete yards should be cleaned and brushed periodically. If a working race is required, it should be just wide enough to take one sheep, the width varying according to the breed. Then at the end a drafting gate may be installed, allowing entrance to one of two pens, and which is useful for running off certain sheep without having to catch them.

Stone Walls

In a suitable area, basic knowledge of stone walling is an asset. Walls are built up quite steep slopes and on ground so rocky that no stake will penetrate. They are fireproof and long-lasting, require no imported or bought materials, and give better shelter from wind than a solid wall, and the air filtering through dries a fleece.

First step in building is to cut out the top soil, and place flat, square-ended stones to make a base slightly wider than the wall is to be. A frame at either end of the section, joined with strings gives a guide for dimensions, and each layer of stones is kept flat, the middle or hearting being filled with small, hard stones, *not* with any old rubbish that will disintegrate in a few years.

Corners should be square, but along the length of the wall stones of various shapes and sizes must be incorporated. A pointed hammer is useful in trimming stones, and heavy gloves prevent damage to hands. The two cheeks of the wall are built separately, but every two feet or so in height they must be joined by 'throughs' or 'thruffs', large flat stones which either fit the wall exactly or project an inch or so on either side. As the top narrows, a row of coping stones, on their ends, completes the work.

4

LOWLAND AND SEMI-UPLAND BREEDS

Some sixty breeds and crosses are listed in the National Sheep Association's admirable book *British Sheep.* This number seems a veritable jungle to the first-time buyer. It need not be so, and we will try to simplify initial choice. Sheep are relatively inexpensive to buy, and the main thing is to *start.* In a year, having learnt more than you thought possible, a change of breed can be made if desired.

We will shorten the list by omitting breeds used specifically for ram breeding. They are crossed with ewes of other breeds, and no one keeps them pure for any other purpose. As the self-sufficiency farmer is concerned with wool, and possibly meat, rather than stock for sale, the Border Leicester, Blue-faced Leicester, Teeswater and Wensleydale may be discounted. The novice may buy lambs sired by one of these breeds, or he may buy such a ram to put to his own ewes of another breed, but he will *not* be breeding them pure. Their production is a highly specialized, professional game.

Other breeds may be roughly classified as lowland, semi-upland, hill and rare breeds, bearing in mind that such divisions are arbitrary and not too rigid. For more details of all these breeds see the Breed Society List.

LOWLAND BREEDS

Colbred

This is prolific and deep milking, not difficult to manage, but available from a limited number of breeders.

South-West Breeds

The south-west has its own breeds, very suitable for anyone in or near that area. They are not yet common in other districts. The Devon

The Dorset/Lonk is a more unusual combination, but its quality for breeding leaves little to be desired. Note the type of pen, which though temporary is strongly set up to hold any required sheep.

Closewool, Devon and Cornwall Longwool, and Dartmoor all give mutton and wool in abundance.

Devon Closewool

The Devon Closewool is a medium-sized white-faced sheep, without horns and with excellent wool. An adult ewe weighs 130-140 lb, and a lambing percentage of 150 is claimed. The lambing percentage is the number of lambs reared compared with the number of ewes put to the ram. Thus, if you put six ewes to the ram in October and have nine lambs to tail the following May, the lambing percentage is 150. (This may also be expressed as 'a lamb and a half'. Sometimes the geld or barren ewes are not counted in the average and lamb losses are ignored, which gives a completely misleading figure.)

The Devon Closewool is one of the few breeds that may be shorn as lambs, when yield is 2½ lb (about 1kg). A yearling ram yields 13 lb (6kg) of wool, a yearling ewe 8 lb (3¾kg), with staple length 5 inches

(13cm) and a Bradford wool count of 46-50 (see Chapter 10). The wool is used for blending, and for hosiery, fabrics and tweeds. It is primarily a grassland sheep.

Devon and Cornwall Longwool

At 180 lb (81½kg) liveweight, the Devon and Cornwall Longwool is heavier. Its fleece has a count of 32-36. For those wanting early fat lamb coupled with a good fleece, this breed deserves consideration.

Dartmoor

Of similar size is the Dartmoor, with 140 per cent lambing, and hardiness brought about by the breezes on the environs of the Moor. Fleece weights of 26 lb (12kg) from yearling rams and 17 lb (7½kg) from yearling ewes are claimed, with a staple length of 9 inches (23cm) and a count of 34-38. In commerce the wool is made into blankets, serges and carpets. The improved Dartmoor is an offshoot, more suited to higher conditions.

Any of these breeds might suit on lower ground in or near the south-west. They could do equally well elsewhere, but beginners must avoid pioneering. To buy sheep away from their own district adds enormously to travelling expenses, and makes sales more difficult through the breed being unfamiliar to buyers. Newcomers to sheep have a penchant for believing they are God's gift to the backward farmers (who have learnt several generations of sheep lore but don't often write it down). After a few expensive mistakes and disastrous losses the same beginners become ultra-traditional! Don't add to their numbers. Play safe, and stick to stock proven in the area.

Leicester and Lincoln

Two other heavy-fleeced lowland breeds that may be considered together are the Leicester and the Lincoln. Both are big, quiet sheep, hornless and white-faced. Their fleeces weigh about a stone (6kg) on average throughout the flocks, yearlings of both breeds being capable of exceeding 20 lb (9kg). Wool from the Leicester is in locks the thickness of a finger, while the Lincoln claims the longest staple of any sheep in the world. It is found mainly in its native county, while the Leicester is now concentrated in the North and East Ridings of Yorkshire. Both breeds are tough, but have large and demanding appetites.

Border Leicester sheep.

Romney

Another hardy lowland breed is the Romney, developed on the exposed Marsh of that name, where lambing seldom starts before April. This is a big sheep, with ewes weighing 160 lb (72$\frac{1}{2}$kg) and lambing 150 per cent, and yielding 10 lb (4$\frac{1}{2}$kg) of wool per fleece from recorded flocks. Staple length is 5$\frac{1}{2}$ inches (14cm), with a count of 48-52. The Romney has sound feet and is accustomed to high-density stocking.

Wiltshire Horn

A breed of no interest to spinners and weavers is the Wiltshire Horn. This is a sheep with no fleece and a skin like horses or cattle. Its mutton qualities are excellent and its protagonists claim that it is a very trouble-free sheep to manage. If you want meat only, here is a fine breed, but you won't make many pullovers from a Wiltshire. A ram could be used on wool-productive ewes where the lambs are wanted simply for eating.

Down Breeds

The Down breeds and the Ryeland are similarly used as sires of meat lambs. They are seldom prolific enough to be considered on their own. The sheep man must know about them, however, for Down crosses constitute the larger part of fat lamb on our tables. They are the Southdown, Suffolk, Hampshire, Dorset Down, Shropshire, and Oxford. The Southdown is the smallest and the Oxford is the biggest of the Down breeds.

Southdown and Hampshire

For neat, quality lambs the Southdown is unsurpassed, while for weight the Oxford is outstanding. It must not be used on small ewes, however, or lambing difficulties may ensue. In general, the halfbreds – Masham, Mule, Welsh Halfbred and Scottish Halfbred are suitable for crossing with the Oxford; the pure hill breeds are not.

Though the Hampshire is an excellent mutton sheep, watch that the ram is not too wide in the shoulder, for that creates lambing problems. The Dorset Down has been systematically bred as a fat lamb sire for generations. It is most readily available in the south and at Kelso Ram Sales, Roxburghshire.

Suffolk and Shropshire

Most popular of all is the Suffolk. Its lambs out of any ewe have smart black or brown heads, are strong and active, and grow quickly into quality meat. The short, white, dense fleece of the pure Suffolk produces a hosiery type of wool. Fleece weight is 7 lb (3kg) per ewe, and the breed is quite prolific for a Down, with recorded percentages of 130 over many sheep. In general, the Downs do not breed great quantities of lambs, which is the prime reason for discounting them as pure breeds for the novice. A great attribute of the Suffolk, and passed on to some extent in its crosses, is its capacity for breeding early in the year.

The Shropshire is an exception to low prolificacy in Downs. These medium-weight sheep lamb 150 to 175 per cent, average 8 to 10 lb ($3\frac{1}{2}$-$4\frac{1}{2}$kg) wool over the flock, and are excellent pasture sheep. They are adaptable and deserve to be better known. The Shropshire flock book of 1883 was the earliest published in the sheep world.

Ryeland

I like the little Ryeland. One of our oldest breeds, it originated in Herefordshire 800 years ago, and like the Downs it is a 'crossing' sheep. I returned one by rail after using it to serve my ewes, and it trotted across York station on a collar and lead like a dog. It was a great creator of conversations. Ryeland wool is soft to handle and of good colour. Its strength is sound and its degree of springiness unsurpassed. There is a total absence of black fibres, and the wool is popular for hosiery and knitting yarns.

GRASSLAND AND SEMI-UPLAND BREEDS

Beulah Speckled-face

In the Eppynt and Llanafan districts of Wales, a speckle-faced sheep has been known for over a hundred years without introduction of female blood from outside. Known as the Beulah Speckled-face, this breed is bigger, more productive and less hardy than the true mountain sheep. No one sheep combines all the virtues, but the Beulah fits into a very handy slot with its ability to move between enclosed fields and the better hill pastures, and rear twins while doing so.

This type of management makes it less wild than some of its more active brethren. Lambing percentages can be 175 and even 200; fleeces are variable but weigh from 3 lb to 5 lb (1-2kg), the heavier ones tending to be coarser. A count of 50 to 58 underlines the variation, finer wools being used for knitting, while the coarser types are suitable for carpets and rugs.

Clun Forest and Kerry Hill

Two breeds of the Welsh borders, the Clun Forest and the Kerry Hill, are worthy of attention. Both are suitable for self-replenishing flocks on grassland, and are generally quiet sheep. The Clun was a regular winner in Farmer and Stockbreeder Lambing Competitions, averaging 'twins all round' in the very best flocks, and a 175 per cent lambing average is attainable with care. The fleece holds a spinning count of 58, and consistency from neck to britch is a feature. Ladies' knitwear, spinning yarns and all types of felting are made from Clun wool.

There is an active breed society, and classes at most major shows.

Black with white face markings, and sometimes black on the knees, the Kerry Hill is easily recognizable. Like the Clun, it is of medium size, and may be bred pure or crossed with a Down ram. Its white, dense fleece is classed with the Downs breeds for marketing purposes, and has a 56-48 count.

Rough Fells are quieter than most hill breeds. They also have heavy fleeces with a staple of 18 inches in the longest parts. Each individual fleece of any breed varies so much in its different parts that average staple lengths tend to be meaningless.

Hill Radnor and Llanwenog

The Hill Radnor is another grassland breed. It is brown-faced, with white, dense, fine wool yielding ewe fleeces of 4½ lb (2kg) from mountain grazings and 6½ lb (3kg) from lowland pasture. Lambing percentage ranges from 120 to 165 according to environment. Breeders claim that the Hill Radnor holds its fleece well before clipping, an important point for the wool enthusiast, to whom a sheep that sheds its wool in large lumps in April and May is anathema.

No one farming sheep in Wales can complain of lack of breed choice, the Llanwenog being extra highly prolific. This breed is of medium size, ewed weighing about 130 lb (59kg), and is found chiefly in West Wales, particularly in Cardiganshire and Carmarthenshire. A lambing average of 230 has been recorded, triplets being commonplace. The fleece is of medium staple, high grade, and 4 lb to 5 lb (2-3kg) in weight. Llanwenogs thrive at 1000 feet in their native district.

Lleyn

Another highly productive Welsh sheep is the Lleyn. This is found in a favoured area of coastline opposite Anglesey, and is very highly prolific. 'If it is so good, why has it not spread further?' is a reasonable question. The answer is in the hundreds of thousands of little Welsh ewes that separate its territory from the rest of Britain. They formed a buffer which buyers need not bother to cross. The Lleyn is now spreading further afield, and is worth a second look.

East Friesian

The East Friesian is a large, deep-milking Continental breed. It is a most useful smallholder's sheep, but numbers in Britain are so few that it is at a premium. Its wool is good and its fleece heavy. Don't be misled, though, into thinking that a change of breed will act like magic, and bring instant success. All our breeds are useful, or they would not have survived. Any breed is only as good as its management.

5
HILL BREEDS

We must now consider the hill breeds. Remember that if you are in the lowlands, draft ewes (see Glossary) from these breeds may still be very suitable. The converse is not true; lowland breeds cannot be made to thrive on the hills, especially in winter.

Exmoor Horn and White-faced Dartmoor

Beginning again in the south-west, the Exmoor Horn provides a better than average carcass for a hill sheep. It thrives on both moorland and grass, averaging 135 per cent lambs on the harder conditions, and considerably more lower down. The Exmoor Horn is often successful in carcass competitions, while it is one of the few hill breeds to produce fine wool. With a count of 48-52, the fleece has sound staple $3\frac{3}{4}$ to 4 inches (9-10cm) long, and is sought out by hosiery and tweed manufacturers. The wool price is good and the yield $6\frac{1}{2}$ to $7\frac{1}{2}$ lb (3-4kg) per ewe. In or near the south-west it is a proposition.

The White-faced Dartmoor grazes up to 2000 feet in summer, and most flocks are kept on the Moor from May to December. The white wool is of good staple, with a fairly strong crimp, and ewes average 12 to 16 lb ($5\frac{1}{2}$-$7\frac{1}{4}$kg).

Welsh Mountain Sheep

There are lots of different types of Welsh Mountain sheep, varying in size from small to medium-small, and in face markings from white to tan, or an admixture. They form the sound basis of the Welsh hill sheep farming industry, yet adapt to intensive conditions when the need arises. Bred pure, or crossed with a Suffolk for meat or a Border Leicester to produce the Welsh Halfbred, they must be considered by anyone within range of the Principality.

Males in the Welsh Mountain are horned, females hornless.

Lambing percentage is from 100 in small ewes or hard conditions, to 150 in the wealthier drafts. The meat makes excellent eating, being tender and of superlative flavour. There is active co-operation between Welsh farmers and colleges, and work is being done to improve fleece quality of the Welsh Mountain. The wool gives soft, durable knitting yarn, with the best fleeces weighing 4 to 5 lb (2-3kg), but smaller ones only half that. Almost half a million draft ewes leave the hills of Wales each autumn. If you fancy a pen of them, do make sure that they are the size and type you really want, as there is no such thing as a standard Welsh Mountain sheep.

Black Welsh Mountain

If the self-sufficiency enthusiast wants black wool for any purpose, he need not dye it. The Black Welsh Mountain provides it in abundance. This sheep originated from Welsh Mountain, and by simply picking out the odd black ones and inter-breeding, a distinct breed has been set up, with its flock book over half a century old.

In the Middle Ages the black wool, known as Cochddu, was much sought after, and today the spun fleece may be used pure, or with good quality white wool to make a most pleasing cloth of checks and patterns. The grade is 54-56 demi-lustre, 1 lb ($\frac{1}{2}$kg) best wool makes 1 yard (just under a metre) of cloth, and fleece weight is 4 lb (2kg). Prolificacy is high for a small sheep, but watch the bottom as well as the top of your fences. Black Welsh Mountains are great creepers!

Derbyshire Gritstones and Lonks

For anyone keeping sheep from Derbyshire to the Scottish border, a whole selection of dark-faced hill sheep is available. They are mostly horned in both sexes, but the Derbyshire Gritstone is polled. It has come into national prominence lately through attempts to 'breed the horns off' the Blackface. The Gritstone is a true hill breed from the bleak southern Pennines. It has a mottled face and a strong carriage, and wool claimed to be the best from any black-faced hill breed. The fleece is dense, white and weather-resisting, lustrous and uniform. Clip is 5 to 6 lb ($2\frac{1}{4}$-3kg), staple length 6 to 8 inches (16-20cm) and the quality count of 56 makes it suitable for high-class knitwear.

The Lonk inhabits a similar inhospitable clime. It has strong and sweeping horns in the ram (ideal for stick dressing) and strong and flat in the ewe. Large for a hill breed, it can withstand damp soil; one flock I visited suffered two hundred rainy days a year, and as mists and industrial fogs are common in its habitat, the breed has developed strong self-preservation instincts, for it often lives and lambs alone. Shepherding in mist can do more harm than good. Lonks often figure

The Herdwick is one of the cheaper breeds to buy, and gives wool of different colours according to its age

in Mountain Wool classes, and stock ewes average $7\frac{1}{2}$ lb ($3\frac{1}{4}$kg) fleeces; hoggs a pound more. The rams will clip up to 15 lb (7kg), and the grade is 56-44. This is an interesting sheep, often travelling around in family parties with last year's lamb trotting alongside this year's. 'One shake and it's dry' is one farmer's description of a fleece that seldom enjoys much sun.

Rough Fells, Dalesbred and Swaledale

Rough Fells give the longest staple, 8 inches (20cm), of any hill breed. Their fleece weight is good at 5 lb ($2\frac{1}{4}$kg), the wool being used for carpets, rugs and mattresses. They are claimed to be the only hill breed that may be worked without a dog! The Kendal area of Cumbria is their home.

The other breed of the north-west is the Dalesbred. It is ideal for breeding the popular Masham when put to a Teeswater or Wensleydale ram, and has a durable, springy fleece of 4 to 6 lb ($1\frac{3}{4}$-3kg), grading 40-32 and used for carpet making and Scottish tweeds. The staple is long and curly.

Fleece weight in the Swaledale is generally lower, that most popular hill breed of the north being valued chiefly for its ranging and milking abilities. It is primarily sought after to breed the Mule ewe lamb, an expensive species capable of producing fat lambs on a big scale. The beginner must not delude himself into thinking that any old Swaledale ewe put to the first Blue-faced Leicester tup available will breed a ewe lamb worth £40 or £50. She won't. That game is in the hands of very skilled and experienced professionals. But the Swaledale can be a most useful sheep, and there are enough draft ewes of unfashionable colours that will breed lambs just as good for the deep freeze.

Bradford Institute of Technology has produced light-weight fabrics of hard wearing quality Swaledale wool for both men's and women's wear. I have a Swaledale suit that gives excellent service. Knitting wools also come from the Swaledale, ewes of which may be found at any dales sale from the Lakeland hills to the North Sea coast.

Herdwicks

The highest Lakeland hills are the home of the Herdwick, that little sheep of legend, black and grey in colour. Herdwick lambs are black when born, going gradually lighter until they are almost white in old age. This allows for a whole range of natural colours within the one breed.

Herdwicks live to a great age. Usually horned in the male, they are polled in the female. Though they can and do scale rocks dangerous to human climbers, the breed is remarkably quiet when brought to low ground. They don't have a big lambing average, but one good lamb per ewe on heavily stocked lowland grass would be a worthwhile venture. It was something I wanted to do during my farming days, for I was always fond of the little Herdwicks and their coats of 'John Peel grey'. Some excellent Herdwick tweeds and hats are available, and Herdwick mutton was used at Queen Elizabeth II's Coronation feast.

A breed of the uplands. Hill Cheviots in their native haunts. They may also be kept on low ground, and provide a fine white wool.

Scottish Blackfaces and Cheviots

North of Herdwick and Swaledale country, the Scottish Blackfaces begin. They are the most numerous breed in Britain. Fleeces vary according to district and climate, some short or medium fine ones from Galloway have a count of 40, and a weight of 3½ to 4 lb (1½-2kg). Others are stronger, weighing 4 to 6 lb (2-3kg). This strong wool is the finest mattress filling there is; demand in Italy is good, so why not use it at home? The finer wools are for hard-wearing clothes. Male and female are horned.

The other Scottish hill sheep is the Cheviot. There are now two breeds, the North country being a bigger animal more suited to marginal pastures. It is a prolific and gentle mother, though rather prone to udder troubles and to shedding its fleece. The wool is used in the tweed industry, and has a Bradford count of 50-56 and a staple length of 5 inches (13cm). It is white, and free from kemp.

The 'Northie's' cousin is the Hill Cheviot. It is a dual purpose sheep with white, dense, lustrous wool of uniform quality, free from kemp. Fleece weight is 5 to 6 lb ($2\frac{1}{4}$-3kg). Crossed with a Suffolk, the mutton is extra good. Bred pure, the Cheviot is a 'spring' sheep, making an ideal weight in March and April, when fresh lamb is not plentiful. The Western-type Cheviot is often horned in the male, usually polled in the East, the ewes of both types being hornless. This is a grassland breed; it will thrive at considerable elevation, but not so well on heather as does the Blackface.

Welsh Halfbred ewes with Suffolk-cross lambs on a north country holding. The Welsh Halfbred is readily available in its home country where its sales are well organized.

Dorset Horn and Poll Dorset

Before dealing with the crosses and rare breeds, there are two others with special points to consider. The Dorset Horn and the Poll Dorset will breed twice a year. In practice, three times in two years is nearer the mark, the important feature being that they will take the ram at any season, and this is unique among British breeds. Dorset Horn milk has been used for cheese making. The wool is among the whitest of British wools, with a count of 54-58, while autumn-born lambs will clip $2\frac{1}{2}$ to 3 lb (1-2kg) the next summer. Lowland or semi-upland conditions are required.

Jacobs

The Jacob deserves a place of its own among self-sufficiency advocates. The sheep is small and prolific, with mutli-coloured wool that makes up into a wonderful array of garments, from thick tweed to light-weight dresses. Jacob mutton has a distinctive flavour. A major point in the breed's favour is its organization, which caters specially for small flock owners and those interested in spinning and making up the wool. Stands are manned (or more usually womanned) at all major shows. A sheep skin rug from a Jacob can be of a pretty pattern.

Crossbreds

Though we have dealt with the purebreds at length, the majority of sheep in Britain are crossbreds. This is no mere accident. Each cross is designed to incorporate the good points of both parents, and usually does. The proven system of 'sheep stratification' begins on the hills, from where draft ewes are sold each autumn. These are mated to a prolific longwool ram and resulting ewe lambs have milking potential, mothering qualities (the two invariably go together) quick growth rate and high prolificacy. Mated to Down rams (see page 35) they give fat lambs of the desired conformation in large numbers.

The most popular halfbreds are the *Welsh Halfbred, Masham, Mule, Romney Halfbred* and *Scottish Halfbred.* The Welsh is the smallest and the Scottish the largest. All are excellent, when making a living out of sheep. Yet they may not suit the small farmer wishing to breed his own replacements and improve his strain. Breeding stock must be brought in, for on limited acreage it is impractical to keep the parent stock *and* the halfbred.

Rare Breeds

Finally we have the rare breeds. Their advantages to the small keeper are two-fold. No one else is keeping them in large numbers, and by joining the Rare Breeds Survival Trust at Winkleigh, Devon, the new member is assured of a friendly welcome from like-minded individuals backed by a professional information service. There is also the satisfaction of keeping alive a breed, once widespread, that so declined in numbers through changing economic circumstances that it was in danger of extinction.

Cotswolds and Island Sheep

The Cotswold is the largest of these breeds. They are a big, hornless, long-woolled, white-faced sheep. They need good food and are not too prolific, but they yield 15 lb ($6\frac{3}{4}$kg) of 40's count wool from young ewes, with a 12 to 14 in. (31-36cm) staple of beautiful regularity.

A group of island sheep deserve serious consideration. The *Shetland, St Kilda, Manx Loghtan, North Ronaldsay* or *Orkney*, *Portland* and *Soay* all have special characteristics. They are small, and less 'developed' than sheep so far considered. Yet their wool may be admirable; both the Manx Loghtan and the St Kilda are multi-horned, the North Ronaldsay subsists entirely on seaweed except at lambing, the Portland makes delicious eating. Another rare breed is the large *White-faced Woodland*, a hill breed centred round Penistone, West Riding.

Merinos

Although Merinos exist by the million overseas, this wool-producing breed is found mainly on research establishments in Britain. It yields fine wool par excellence, but it cannot be recommended to the novice.

Continental Imports and Others

Recent Continental imports are the *Oldenburg* and the *Texel*. The former is a very quiet sheep used as a 'crossing' ram on hill ewes, the latter a meat ram. The *Finn* is variously regarded as the biggest disaster inflicted on British sheep since the Second World War or as a means of boosting prolificacy. It cannot be recommended to anyone. The *Cambridge* is an amalgam of very prolific sheep, and is still in the formative stage.

6

FLUSHING AND TUPPING

The shepherd's year traditionally begins with tupping, or mating, which for most British breeds is in the autumn. It is more practical to anticipate this date, and consider management of males and females before tupping.

The most likely tupping dates are between late July and late November, and for the majority of flocks it is October and early November, to give March and April lambing. Choice of this date is critical, and must be determined well in advance. Find the local custom, evolved over decades of experience by professionals, and above all don't try to be clever and lamb early on the evidence of one gentle spring. Sheep almost shepherd themselves when conditions are right. Adverse seasons are those that call for careful shepherding.

Gestation period of the ewe is about five months. Early maturing breeds tend to have a shorter gestation period – Ryder and Stephenson quote 144 days for Southdown and Dorset Horn, while the Merino is 151 days – and hill breeds carry their lambs rather longer than lowland. If the ram is turned among Swaledale ewes on Bonfire Night, 5 November, the first lambs may be expected on All Fools Day, 1 April. These are handy dates to remember and work around.

Self-sufficiency sheep people are seldom interested in early lamb for butchering, for the only point of lambing early in the year is to catch the spring market with lambs sold fat and straight off their dams' milk. Unless that is the aim, avoid the exercise. It adds to cost because both mother and offspring require extra concentrate feeding, since there will be precious little grass at that time of year.

The best practical lambing date is the one when grass may reasonably be expected to grow. British springs vary so much that you cannot be right every year, and lambs will be born over at least a three week period, but March is the usual month for most lowland situations, with dales flocks starting in early April and the true hill flocks being held back until late that month. One very experienced County Durham hill farmer said that if he could lamb all his several hundred Swaledales on one date, he would choose May Day. In practice he timed tupping so that the first lambs were dropped on 25 April.

Having decided on which date the rams will be turned in to the females, we now consider the ewes' management. The aim is to have them in good but still rising condition at tupping. At one time it was thought best to have them really lean some six to eight weeks before tupping, and then build them up. This has been disproved by New Zealand work, which shows that, provided the ewes are well fleshed at tupping there is no point in deliberately putting them on short commons beforehand.

The object of this period of 'flushing' the ewes is to cause them to shed the maximum number of eggs. If a ewe sheds but one egg she can have but one lamb, but she is likely to conceive twins or triplets if more eggs are shed. Fecundity of the sheep tends to be higher in the longwool breeds than in the Downs, which in their turn produce more lambs than hill breeds kept in their natural environment.

Flushing

Of all the different ways of 'flushing' the ewes, the easiest and cheapest is to use a field of fogg or aftermath. This is the regrowth after the meadow has been cut for hay. After the hay crop has been removed in June or July, the area is given a top-dressing of about 30 units of nitrogen per acre, according to the season and the need. In a wet, 'growy' summer, there may be little point in spending money on bag nitrogen. This saved grass is also excellent for fattening lambs.

Another method is to grow a crop specially for the ewes. Mustard and rape, fodder radish and 100-day turnips are all possibilities. These are all quick-growing crops which in farm practice usually follow an early-harvested cereal in August, but on the smallholding they may be sown after any summer ground clearance.

Using Concentrates

If neither 'green crop', as these last-named are termed, nor new grass is available, ewes may be helped by a small concentrate feed. One quarter or half a pound (100/225g) daily to a medium-sized ewe certainly helps, and though it is expensive in view of today's corn prices, quite a small increase in the lambing percentage will pay for it. One of the best stockmen in North Yorkshire practises nut feeding every autumn just before the ewes go to the ram, and seldom falls far short of the ideal of 'twins all round' from his Masham ewes.

Watching for Lameness

During this time watch the flock for any sign of lameness. Foot rot can spread rapidly through a batch of sheep in autumn's wet conditions, and afflicted sheep lose flesh rapidly, the opposite of what we seek. Another reason for foot care now is to avoid undue handling when the ewes are safely in-lamb.

Nor must the rams be neglected. They often are, until actually needed. My vet tells me that every autumn he has a stream of panic-stricken calls from farmers who, busy with harvest or the hundred-and-one late summer jobs on a hill farm, have suddenly discovered that their rams are stumbling about with feet like footballs. This is sheer bad shepherding, inexcusable though understandable when today's small staffs are faced with too many jobs, and it is a mistake that the self-sufficiency advocate with only a few sheep simply dare not make.

If the slightest hint of lameness occurs, catch the sheep, inspect the feet, use the knife where necessary and give a squirt of foot rot spray into *each* hoof. Rams have a regular and infuriating habit of walking perfectly soundly throughout the ten months of the year when they don't 'earn salt', and going lame just when their services are required.

Fighting

Another problem among rams is fighting. If you have only one, leave him with an old ewe for company while the rest are being 'flushed'. With two or more rams, there is constant danger of fighting as the days shorten. Don't be misled by the sight of a bunch of rams lying about contentedly all summer, like so many old gentlemen on a park bench. With lengthening darkness the hormones controlling libido in the ram come into play, and they take on entirely different characters. One severe blow in the right place is sufficient to lame or even kill another ram.

The biggest danger of this happening is when rams are evenly matched. A ram lamb and an older one seldom fight; the real trouble arises among mature animals each of which thinks he is the boss. To circumvent accidents, one plan practised among breeders of horned sheep is to couple the rams by means of a length of swivelled chain attached to a horn of each. I don't like this. On one occasion the weaker of a coupled pair was not allowed to the water trough by his mate. The shepherd did not realize what was happening, and when the sheep was finally released he drank so much that he died.

Former Shepherd of the Year Ray Dent looks over his selection of Swaledale rams. Though the self-sufficer will probably need but one ram its choice is vital for it will be responsible for half the characteristics of the next lamb crop.

Housing Rams

A better plan is to house the rams, feeding them on hay and cabbages and perhaps a little corn. A leading Northumbrian hill farmer tells me that as soon as the leaves begin to fall he takes his rams into a building, as they cannot be trusted from that time. If rams are housed in a strong building, you do at least know where they are. Few things are more frustrating than to have carefully-laid breeding plans set at nought by the wrong ram getting to the wrong ewes at the wrong time, be it your own or a neighbour's.

If rams are in a paddock, fences must be very strong, for they will break out at this season although they have lived happily in the same ground all summer. Rams wandering about the countryside cause a lot of ill-feeling and strained relations with neighbours. One ram was reputed to have swum the Bristol Channel where it was a mile wide, tupped a number of ewes and returned home again, all in one night, so don't rely too much on natural barriers!

Beginners are often put off by the complications of lambing time. Yet a healthy flock of mature ewes will largely take care of itself at that season. If things go wrong at tupping, there will be no lambing, so the autumn season is at least as important. Start with a proven ram of middle years. By proven I mean one that left healthy lambs the previous year. Though there is no guarantee that he will do the same again, the odds against failure are lowered. Avoid ram lambs or aged rams of more than four-shear. They are for the skilled man, not the novice.

Learn to Look

You may learn a great deal about sheep simply through watching the flock at tupping. If the Victorian farmer asked his shepherd of the folded flock what he was doing, when he was apparently doing nothing, the stock reply was: 'Oi be watching your sheep, maister!' Only through constant observation may we understand, and the ram's aggressive male characteristics, his pawing of the ground, his sniffing at each ewe to see whether she is in season, his grunting and rubbing alongside ewes he has just mated, are there for the eyes to see.

Colour Marking

However, you cannot be there all the time, and the way to tell which sheep have been mated is to colour the ram's chest. Ruddle in various colours may be purchased as a powder, mixed with water and smeared on the ram's chest. The bare part of the breast bone is the place to apply the ruddle. It may be done with the ram still standing, using a piece of flat wood, while he eats nuts to keep up his strength. It may be done by turning up the sheep, but that is no light task for a small person and a big ram anxious to be back among his ewes.

The other method of colour marking is by means of a ram harness with coloured chalk attached. Once attached, the harness remains *in situ* throughout tupping, the chalk on the breast plate being changed as desired. The harness had advantages: it may save time as there is no need to catch and ruddle the ram daily, and a wide range of coloured chalks is available. On the other hand, it may lead to lazy shepherding. It is an attempt to 'mechanize', and mechanization and good husbandry do not always go hand in hand.

Because the ram does not have to be caught daily, there is the temptation not to bother to feed him daily, especially if he is an awkward customer as rams tend to be. In wet weather especially, the harness may chafe under the armpits and cause lameness and a disinclination to mate. If this happens, remove the harness, allow the sores to heal and resort to ruddle paste. Sometimes a new chalk block has such a fine, hard finish that it does not mark properly. The simple remedy is to rough up the surface with a sharp tool.

Heat Periods

During the breeding season each ewe comes into heat every sixteen or seventeen days. Hill breeds, Swaledale or Scottish Blackface, may have cycles half a day shorter than the Down breeds, and a day shorter than Merinos, but there is no practical difference for our purpose. The duration of heat is some thirty to thirty-six hours, ovulation taking place towards the end of the period. If all ewes conceive on their first heat after introducing the rams, lambing is completed within sixteen to seventeen days of the first lamb being dropped, and this is the shepherd's objective.

In practice, in a healthy flock, there are usually some ewes that conceive 'the second time round', odd ones later still, and a few not at all. Yet in a small flock, every ewe is in constant reach of the ram, and geld or barren ewes should seldom occur. If they do, they can be sold fat in spring when prices are high, or given another chance, but no more.

The coloured chalk or ruddle indicates which ewes have been mounted by the ram. Change the colour after a week if you wish, so that you know which ewes are likely to lamb in the first week, or simply note down ear numbers or names as each ewe is marked. The crucial time comes after the sixteenth day, when a fresh colour *must* be used.

Infertile Rams

The sequence of colours used must be such that each is visible when superimposed on its predecessor. Red is usually the first one, followed by green or blue. After the sixteenth day, watch carefully. If all ewes already coloured are being marked again, suspect the worst. It can only mean that the ram is potent but infertile. Procure another ram immediately; it will probably be easy to borrow one, for shepherds sympathize with each other in such situations. If you have twenty ewes, and the first four or five return to service, that is conclusive enough that something is wrong.

Maximum Fertility Times

We have discussed the ewe's breeding cycle, but there is a further important point. Breeds tend to maximum fertility in mid-breeding season. An upland breed might have late September to February as the period in which its ewes come into heat. Only a few would conceive very early or very late, and they would probably lamb singles. The highest lambing percentage comes from November matings.

We know more about fertility in the ewe than in the ram. Only recently have the great differences between breeds and individual rams been highlighted. Most Suffolk rams work happily enough in August when a hill ram put to the same ewes might be totally disinterested. A Northamptonshire vet found that 10 per cent of rams tested were infertile, and 10 per cent sub-fertile, and that much more attention could be given to this point.

Preferences

Sometimes a ram develops a passion for a particular ewe, does not leave her side and continually remates her. This may be a touching sight, but does nothing to get the rest of the flock in-lamb. When it becomes obvious what is happening, remove the fortunate ewe for a day until she goes off heat. Another trouble that arises in mixed flocks is that rams tend to prefer 'succulent' ewes like Down crosses to

'Twiggies' like the Welsh. These problems do not occur when all ewes are of the same type or breed, regardless of what it is.

With common sense and observation, even the novice shepherd can get through tupping time with good results, and be all set towards a successful lambing. One ram can mate an enormous number of ewes, and generally recommended rates are twenty or thirty ewes for a ram lamb, forty to fifty for a shearling and fifty to seventy for a mature ram. Close relationship rather than numbers is likely to be the small farmer's limiting factor. Ewes should not be mated to their own sires, though daughter to grandsire is generally a safe mating. A simple solution is to change the ram every two years, but where pedigree breeding and flock improvement is the aim, a fairly high proportion of males to females is needed.

7
WINTER, SPRING AND SUMMER

After tupping comes a quiet time in the shepherd's year. All ewes are, we hope, safely in-lamb, but lambing is some three months away. During early gestation the unborn lamb makes little demand on its dam, but be ever mindful of Dr Allan Fraser's dictum: 'The unborn lamb is out of sight for five months; let it not even for a single day be out of mind.'

This view is reinforced by work at Great House Experimental Husbandry Farm, high on the Pennines near Rossendale, Lancashire. Director Mr Harry Mudd showed that resorption of the foetus can occur at any time up to ninety days after conception, and is often associated with periods of stress. These stresses are usually adverse weather or low nutrition, or especially a combination of the two. It is natural for sheep, and especially hill sheep, to lay fat on their bodies during summer and autumn, and feed on it during winter, but this must not be carried too far.

Although the high condition of tupping cannot be maintained as winter sets in, the sheep must not be allowed to become thin. To this end, handle gently but regularly – once a week at least – throughout the winter. Bring them into a small pen or building, being careful not to force them through narrow doorways, so that they have no room to dash about. Put your hand firmly on shoulders or loins, and accustom yourself to the depth of fat cover. If it is shrinking rapidly, more food must be given. Don't judge a sheep on sight; its fleece disguises a multitude of shortcomings. Learn to use your hands as a guide. A sheep that lags behind the rest is also supect, but this is less likely to show up in small fields.

Shelter

'Shelter is half meat' is another of those traditional sayings vindicated by modern research. At least half the food eaten by a sheep is used merely to keep it alive, and on an exposed hill the figure might reach 85 per cent. Blaxter at the Hannah Dairy Research Institute, Ayr, showed in 1963 that a light wind of nine mph more than doubles the maintenance requirements of sheep compared with those in still air. Munro (1962) found that wind speed was the greatest climatic factor in driving sheep to shelter.

Shelter is all-important at lambing time. This cosy corner is at 1,000 feet on an otherwise very bleak North Yorkshire hill sheep farm.

If natural shelter is not available, it may be easily supplied on a small scale by wattle hurdles or lengths of stone wall. Lowland farms can be bleak places in winter if hedges or trees are absent.

Dipping

Early in winter a dip is given. Choose a mild day, and the appropriate winter dip (*not* a summer 'fly' dip); and don't forget to inform the local office of the Ministry of Agriculture if compulsory double dipping against scab is operational. In-lamb ewes may be safely dipped provided they are handled with care.

Always follow the maker's instructions regarding strength of dip. With a small flock there is seldom need to 'top up' the bath during dipping, but when this is necessary, the strength required is different from the original. This is because the fleece filters out a proportion of the dip from the water in which it is mixed.

When dipping, handle gently. In most types of bath, lower the sheep in gently, rump first, and ensure that the body is submerged throughout. Dip the head under two or three times by means of a pole. If you have any cuts or skin conditions, cover them so that no dip splashes on. Usual recommended dipping time is one minute per sheep, which seems an age if she struggles! Contact the offices of the Ministry of Agriculture for your area, and find out the position regarding compulsory dipping, and the amount of notification to be given. These compulsory dipping regulations may be varied from time to time; it is your job to keep up to date with them.

The object of dipping is to wet the sheep thoroughly right to the skin, and this is not always easy. Save clothes fit only for the rag-and-bone man for dipping days; you will need them!

Watch for signs of lameness and take appropriate action. Take note of any gaps used by sheep passing from field to field. One thing I cannot abide is a length of barbed wire across a gap that catches the fleece of every sheep passing under it. The same applies to gaps in thorn hedges; if sheep have a passageway, let it be a clear one.

Winter Feeding

Never over-estimate the value of winter pasture. It may appear so good that the sheep refuse to eat hay, but be prepared to feed a few nuts if this happens. An open winter has its troubles, failure to eat sufficient hay being one of them.

If the flock is housed, there is no temptation to rely on natural herbage. Everything must be provided, and this includes access to minerals and water. Sheep drink a surprising amount of water, and care must be taken that their supplies are not frozen.

Good hay and a few roots or remaining green crop suffice for the outdoor flock until six weeks from lambing, mineral licks or powders being available throughout. From six weeks before lambing a balanced concentrate is introduced, at the rate of $\frac{1}{4}$ lb (100g) per head per day for medium-sized sheep. This is gradually stepped up until 1 lb (450g) (or even more among large breeds) is given in the final week or two.

For these last weeks the flock may be divided, or individual sheep fed separately. Some ewes thrive better than others, certain sheep may be carrying twins or triplets, while lambing dates vary. During these last six weeks the unborn lambs are really growing, and they and the surrounding placenta also take up a lot of room. This means that less is available for food, which should become more concentrated.

Care of the young replacement sheep is an important aspect. These Scottish Blackface ewe hoggs have wintered well.

At the Institute of Animal Physiology, Babraham, Cambridge, the flock of registered Cluns averaging 190 lb (86kg) liveweight is restricted to $2\frac{1}{2}$ lb (1kg) total hay and concentrates per day. When first housed they receive most of the ration as hay, but as lambing approaches the hay is cut down and the concentrates raised. The flock under Mr Tom Fulton's managership has won many National Lambing Competitions and achieved a lambing average of 198 per cent.

During this winter period, the non-breeding sheep require regular feed but little other attention. Replacement ewe hoggs, if not in-lamb, and wether sheep need good hay and perhaps a few roots, but little in the way of corn. Their worm drenches, vaccinations and dipping coincides with the rest of the flock. Ewe hoggs are still growing, and feed and shelter to supply this need is required in addition to maintenance.

LAMBING

A week or two before the event, the lambing paddock must be decided on. On a smallholding there may be no choice. Shelter becomes more important than ever, and a building for night lambing and special cases is a blessing. A few small individual pens about six feet by three feet are erected by means of straw bales or hurdles. If these are under cover, so much the better, but if outside they should be under a sheltered wall and may be partly thatched or roofed with corrugated iron.

Equipment

A feeling of confidence is generated by preparing these pens and having the necessary equipment to hand. The list may seem long and comprehensive, and with a small number it will not all be necessary in one year, but is worth building up.

Water
Soap
Towels
Disinfectant
Navel dressing
Lubricant
Heat (lamp or good old-fashioned fireside)
Light
Lamb reviver
Transport for animals to vet
Lambing tool
Castration rings
Tailing rings (one and the same)

Pessaries, antibiotics, syringes and disposable needles are useful, but the person with just a few sheep may prefer to rely on the vet when they are needed. Nor is there any point in having a lambing rope and lambing tool while lacking the skill to use them. Former Shepherd of the Year Mr Ray Dent, Weardale, County Durham, adds waterproof clothing to his list. 'The worse the weather, the more important it is for the shepherd to be out and about', he says.

Pre-Birth Behaviour

When lambing is imminent the ewe becomes uneasy, tries to wander off on her own, appears red and swollen at the vulva and has an udder charged with milk. She bleats uneasily from time to time, and may scrape a bed on her chosen place. Do not be misled into thinking that 'Mother knows best'. Sheep will lamb in the most unsuitable places; an exposed hill end, by the bank of a flooded stream into which her offspring makes its first, and last, move, or on a track carrying farm traffic. The ewe must be driven gently to a more suitable place from which she cannot escape.

Giving Birth

Just before delivery the ewe lies down and pokes her head into the air in very characteristic fashion. The water bag appears first, and then the ewe begins to strain. She may get to her feet and turn round before settling again. The lamb's nose should appear, followed by its front legs. Head and neck should emerge, then the shoulders, after which the whole body slithers to the ground. If you are present, pull the lamb round to its mother's head, checking only that nose and mouth are not covered by membrane. The ewe will do the rest of the cleaning.

Difficult Births

If birth has not taken place half an hour or so after straining begins, an examination may be made. Clean, disinfected hands are essential, and lambing oil's a help. The most usual cause of a difficult lambing is for one or both front legs to be stretched back. Work gently along the foreleg, keeping the finger tips cupped so that the nails do no damage, and slide over the hoof. Draw it slowly and gently forward, cupping the fingers over it, again to prevent damage. Do the same with the second leg if it also was misplaced, then continue to draw both together away from the ewe but in a downwards curve, i.e., towards her hind feet.

Another mispresentation to concern us is the breech presentation. If the hind feet are visible, they may be grasped and the lamb drawn out. If the tail is showing, but no feet, the lamb is jammed against the pelvis and must be pushed slightly back into the womb. Then gently hook a finger round the hock and draw the leg forward.

Sometimes the head appears before the legs, which are laid back and unable to pass through the pelvic entrance. The head must then be pushed back, which some prefer to do with the ewe suspended by her hind legs and her shoulders on the ground. It may be impossible to cup the legs forward without pushing the head back in, and heads swell to alarming sizes through being out too long. In the case of twins or triplets, complications arise through having to decide which head and legs belong to which lamb. But in general, lambing a small number of ewes is a pleasant and exciting occupation not to be spoilt by listing all the things that *might* go wrong.

Numbering Lambs

As the lambs arrive, number them when they are dry. A small jar of marking fluid is carried for this purpose, or you may, if you wish, use a marking stick. Each lamb is numbered or marked in the same way as its dam. Thus, mark blue down the near shoulder for the first lambs, then on the ribs, loin, neck or head, and along the other side. This is a private and temporary code. Choose your own method of numbers, letters or marks; it is immaterial provided you have an easy check on which lamb belongs to which ewe. As births come thick and fast, it is easy to have a mix-up.

Removing New Births

If a small paddock next to the lambing area is available, this should be used to push in ewes and newly born lambs. There is always the chance that some ewe that has not yet lambed will have her maternal instincts so roused by the sight of new youngsters that she will take one off with her. She has no milk, of course, and is a danger to the rest if not watched. Another reason for removing new births is to make pressure less and observation easier among the remaining lambers.

If lambing under cover, pop the new arrivals in a specially prepared small pen. If all is well, the next day they may go into a loose box with a few more ewes and their lambs. Watch all the time that they are 'mothered up'. As you go in, and stir up the lambs, they should rush to their mothers' sides and start suckling straight away. If this happens and the ewe stands proudly over them, all is well. If not, and the ewe bumps away one or both of her lambs, they must be returned to confinement in the single pens.

Bad weather is an invariable feature of at least part of a British lambing time. Here Weardale, County Durham, head shepherd Ray Dent (and dog) ensure that a Swaledale ewe is mothering *both* her twins.

In an ideal lambing situation there is a constant flow of newly-born lambs and their dams from the lambing yard or paddock, into small pens or an adjacent area, then into the main grazing area. If numbers are small and this is impractical, no matter. Individual observation is the key, and the various divisions simply make it easier to observe among growing numbers.

Orphans and Rejected Lambs

The occasion arises when a ewe dies, leaving healthy lambs, or has more lambs than she can cope with, or turns nasty against one or both of her offspring. Restraint is then necessary, and the best means is a yoke that prevents the mother from butting the lambs.

'Lamb adopters' may be bought, but are expensive for small numbers, though not where their use is constant. The basic principle is the same; the ewe cannot move away from the lambs, or attack them. Few creatures are so mean, or so unfuriating, as a practised ewe knocking small lambs for six.

'Mothering on'

This sounds a terribly complicated business, but it isn't. When you wish a ewe to take a lamb not her own, the best time to do so is as she is giving birth. Have the foster lamb ready, and as the water bag appears, break it over the orphan and anoint liberally. Then as the new lamb is born, rub it and its 'cleansings' (afterbirth) around the orphan. Let the ewe lick this foster lamb dry *before* she is given her own lamb. The chances are that she will take both. If she has twins herself, the ploy is defeated, and you must try elsewhere.

Feeding

After safe delivery and perhaps a successful 'mothering on', the shepherd's troubles are just beginning. The next lesson is to distinguish between a sick lamb and a hungry one. The latter is always trotting after the ewe with head outstretched as though seeking a teat, and bleating, while a sick lamb lies in shelter or stands with drooping head. Only in really bad weather is there much difficulty in deciding, for then even the fit lambs stand 'with backs like staples', miserable in the rain. Learn to judge a full stomach, by gently feeling that part on an obviously healthy lamb, and comparing it with the 'clapped' appearance of a youngster short of milk.

Bottle feeding is to be avoided if possible, but may be essential if a ewe dies and no foster mother is available, or for triplets gaining insufficient milk from their dam. In any event, it is essential that a newly born lamb has colostrum either from its dam or another ewe that has just lambed. Colostrum in the first twenty-four hours after birth contains antibodies that protect a youngster against a host of diseases, and its lack greatly lowers the chances of rearing that lamb.

Nanny goats are excellent foster mothers. The lamb is allowed to suck little and often, just as if he were with his own dam. Cow's milk is not good. Some brands of proprietary milk for lambs are far better. Lamb teats as supplied by farm stockists are by no means satisfactory at the start; the hole is usually too small, and must be enlarged, and when possible break in a new teat on an older lamb already highly skilled in the art of sucking.

Castration and Tailing

Castration and tailing are two jobs always done simultaneously. All male lambs not required for breeding should be castrated, and the best method is the rubber ring. Do not, however, attempt castration and tailing at birth as is sometimes advocated. Leave it until the lambs are strong and properly mothered up, i.e., at four or five days old when they are, perhaps, being turned onto their summer pasture. Both operations cause a certain amount of stress, so it is unwise to add more stress to a lamb that has not really established itself with its dam.

The lamb is held by an assistant and the ring is slipped over the scrotum by means of special pliers. Before closing the pliers, both testicles must be in place so that the ring will cut off their blood supply from the rest of the body and the rudimentary teats must be well clear. Sometimes the testicles are in the correct place initially, but if not they must be eased down gently. Then the ring is released in position.

Occasionally one or both testicles cannot be located. In this case do not tail. The long tail is the shepherd's sign that, in a breed where docking is normal practice, castration has not been carried out effectively. The testicles may sometimes be located at a later date. If not, the animal becomes a 'rig', and a nuisance, and is best butchered as soon as possible. If the ram is being kept for breeding, he is tailed in the normal way, and given a distinguishing mark.

Hill breeds are left with a long tail for protection and food reserve, or just the end is taken off. Down sheep are usually docked close to the spine, others two or three finger's breadth away. Slide the ring along the tail by means of the pliers, clear away the thickness of wool and release the ring. Lambs show discomfort by rolling about for a quarter of an hour, after which they feel no pain.

Grazing

The best stimulant of all to ewe milk, and therefore to lamb growth, is fresh grazing. If you possibly can, reserve a paddock from the previous autumn. Keep it clear of other stock, top-dress it with about thirty units of nitrogen in later winter, and turn out the ewes with their lambs as soon as they are settled together.

Minerals must be available to sheep at this time. The twice-yearly innoculation of ewes and lambs takes place in spring and autumn. A lamb's early resistance comes from the ewe, but should be reinforced by the needle at about a month old.

Shearing

Shearing is the next major date in the shepherd's calendar (see Chapter 10). After shearing, the wool must be allowed to grow again for ten days or a fortnight before dipping. If this is not done, the skin is too bare and wool length is insufficient to hold the dip. The summer dip is against the blow fly, which lays its eggs on the animal. The larvae hatch and grow by literally feeding on the animal's flesh. If, in an unusual season, a sheep is 'struck' by the blow fly before dipping, the affected parts must be clipped and dressed with a proprietary fluid.

Steady growth is expected throughout the rest of the summer. The season of highest risk is over, and so the flock owner enjoys the sight of thriving lambs and contented ewes in the sunshine. Sheep love sun on their backs. They do far better in a dry spell even though apparently short of grass, than in wet, cold seasons with lashings of pasture all around them.

8
GRASS AND FODDER

There is an age-old saying 'A sheep's worst enemy is another sheep.' It is a reference to the build-up of internal parasites which results in failure to thrive, and a tendency for the sheep population of any given area to die down to the numbers it can support in good health.

Modern science has helped us overcome these restrictions – to a certain degree. Yet no shepherd can afford to forget them. The various methods of worm control help, but they must be sensibly applied. Above all, the newcomer must not fall into the trap of believing that if he has had one successful year, he is bound to do better still the next time, because of his increasing experience. He won't. If the land had been sheep-free for a number of years, that first season may prove the least complicated.

Sheep are not good grazers on their own. They eat out the finer grasses and herbs, and leave the tough ones. These latter then start to dominate the sward, and in a few years its suitability for grazing sheep has declined considerably. Cattle, with their rougher tongues, are ideal for preserving a balance. But the smallholder may not have cattle. If the pasture can be run over with a mower or forage harvester, as it grows away from the sheep, that is also a useful control. However, the pasture may be too rocky or steep, or the shepherd have no mower available.

In this case, grazing by sheep one year and some other stock in alternate years is a practical compromise. Another help is winter housing, which means that grasses do have some time to clean up during part of the year.

If you must use the same sheep pasture year after year, a strict anthelmintic routine is necessary. The main killer worm is *nematodirus*, which tends to attack lambs in May and June (see

Chapter 9). It is difficult even for experienced farmers to realize the speed with which intestinal worms can multiply; one scientist, Mr Brian Heath, gives the best warning by comparing the rate of rise in the worm graph with the speed of light.

Fertilization

Care is needed when fertilizing pasture for sheep. The aim is a constantly-growing, dense sward, not a long, tufty, open one more likely to suit rough Irish bullocks. 'The sheep needs yesterday's grass, not last week's', is wise counsel. On the Romney Marsh – the most densely stocked sheep pasture in the world – it was said that you ought to be able to fling a florin as far as possible, and find it again. Of course, the Marsh has a high water table, and therefore less liable to suffer from drought, but the principle of tight grazing remains sound wherever you farm.

Early grazing for ewes and lambs is of incalculable benefit. If possible, set aside an area, give it a small dressing of 20-30 units of nitrogen per acre, about mid-February or as soon as the ground begins to warm up. In May and June there is likely to be as much grazing as the sheep can cope with, so ease off the fertilizers except on hay ground. Then in late summer and autumn, further dressings of nitrogen help carry grass growth through a longer period. The first week in September is usually the last date to gain economical response.

Hay

If wool is your aim, hay and grazing are the only crops needed. It is possible, and may be more economic, to buy in the small amount needed. Depending on the district and the length of the winter, from one to three hundredweight per ewe will suffice for the winter. Paradoxically, hill flocks may require the least, particularly if they have heather and rough grazing available. Do open a bale when buying; if it is mouldy, clapped tight together in the centre, brown or lacking a sweet smell, leave it alone.

'Sheep hay should come from a sheep farm' is a saying difficult to carry out in practice. It implies, correctly, that the fodder should be sweet and short and herby, like sheep pasture. But sheep will twist up rustling clover hay, especially if the snow flakes are flying, and I believe that there is definite benefit in feeding hay to upland sheep that was made from a fertile lowland ley. It contains minerals that may be lacking in the normal diet.

Hay varies enormously in price. In a good year, it may be plentiful and cheap. In a wet summer, hay is often of poorer quality, and demand forces up the price. On a comparable food value basis, it should be half the price of barley. When barley is £80 a ton, there is little sense in paying £50 or even £60 per ton for hay. Properly made, the crop lasts two years at least, and race horses are never fed on new hay. So if you have storage space (and money) stock up in a good year for the next two winters. The position to avoid is that of being forced into hay buying in February or March, when prolonged snows have caused greater consumption, and the price goes sky-high. Better to pay £100 a ton than lose sheep through starvation, of course, but best of all keep a good reserve.

In an open winter, your sheep may graze at pasture, apparently content, taking little hay if it is offered. This is a dangerous situation. The herbage may meet the appetite but not the nutritional demands of sheep plus unborn lamb. Handle your ewes every week or so during autumn and winter, accustoming yourself to the feel of their spine and tailhead. In this way any loss of flesh is noticed *before* it becomes too serious. The mass of wool hides a multitude of sins, particularly in a dry time, for a clean, dry fleece always looks well. If the sheep are losing condition too fast, and won't eat hay, you must either fasten them up so that they do take it, or feed concentrates.

Making Hay

The principles of haymaking are to dry the cut grass so that harmful bacteria do not multiply in it. If overdried, the fodder loses food value. If stacked wet or too green, mould and spoilage occur. Silage for sheep is a satisfactory food if of high quality, but quantities of a few tons are wasteful, and the heap is not big enough for proper temperature control. Hence we do not consider silage for the small sheep farmer.

Grass is easiest to make into hay when the pollen on most varieties has formed. This is too late for top quality hay; it may be attempted when the flower heads appear. In practice haytime is decided by the weather, and a hot, settled spell in early or mid-June is a boon to the shepherd. Mowing is done by machine or scythe. You cannot always obtain a contractor's services at the optimum time, especially on a small acreage, so prepare an alternative.

Once the grass has been cut, it dries quickest when exposed to maximum air. No more machines are essential on a small acreage. Our great-grandparents made all their hay by hand and wasted less than we do today. The newly-mown swathe should be immediately shaken out in a circular motion of the hayfork tines. Then, as soon as the sun does a bit of work, turn the swathes back. The task becomes lighter as the grass dries; in the early stages, it is very hard work.

Do not be misled into thinking that a baler is an essential piece of haymaking equipment. It isn't. It is simply an expedient for making a lot of hay with a small labour force. Hay may be made safe before it is fit to bale, if you don't mind the work.

The best hay is made on a tripod or rack. The basic principle is to allow maximum sun and wind combined with protection from the rain. A simple tripod with cross-bars or wires, and a barrel at each corner, is covered with layer upon layer of hay. When the apex is reached, the barrels are removed and the spaces act as vents. A rack or succession of strained wires one above the other does the same job, and is a feature of Norwegian or northern Isles haymaking. The English haycock and the bigger pike are variants, and though neither need equipment, the hay is not made safe quite so quickly – and those lost hours may mean the difference between first-class and moderate hay. Stack under cover or under thatch, and if you are unsure that the hay was dry enough, insert an iron rod overnight. If the end becomes uncomfortably hot, the stack will have to be turned over.

Green Crops

Ground for green crops need not be planted before June. Kale for sheep should be leafy rather than stemmy; those marrowstem kale stalks as thick as your arm are for cattle, not sheep. If the land is clean enough, a broadcast crop is ideal. If not, drilling fairly thickly in two feet (60cm) rows allow inter-row cultivations.

Swedes and turnips are good, but June is quite early enough for sowing, otherwise mildew attacks are invited. New dressings to prevent mildew are worth watching. In recent years the 100-day turnip, sometimes but not always accurately called the Dutch turnip, has come to the fore for autumn sowing. It may even be sown after an early corn crop, its value being in the leaf rather than the bulb.

Other catch crops are mustard and rape, and fodder radish. The former has been proved by time, the latter failed to fulfil early promise owing to its susceptibility to wind damage, which can flatten a tall crop and cause much waste.

Mangels

Mangels are another excellent sheep crop. They are capable of great weights per acre; I have grown them as a boy at the rate of 100 tons an acre, without artificials. Care is needed in feeding them to in-lamb ewes, when 5 lb or 6 lb ($2\frac{1}{4}$-3kg) daily is adequate. To milking ewes they may be fed more or less to appetite, and do not lose their freshness when thrown out to pasture.

For a very small area, seeds may be dibbled 9 or 10 inches (23-25cm) apart in 2 foot (60cm) rows, putting three or four seeds per hole and thinning out all save the strongest. The crop has the advantage that it may be grown year after year on the same area. The best mangels I ever knew were grown on a sewage farm, and received nothing save sludge. Exhibitors from miles around procured their mangels there! Feeding should not start until after Christmas, as the crop is unripe before then, and it must be stored free from frost. It is a great stand-by in a backward spring.

Kale, rape and 100-day turnips are best fed *in situ*. The sheep are let onto a small area daily, or every other day, and made to clean up before moving on to the next section. Wattle hurdles or wire nets are used. The combination of root fibres, droppings and treading action is unexcelled for land improvement, resulting in the sheep's synonym The Golden Hoof.

9
DISEASE PREVENTION

Nothing could be more off-putting to the would-be shepherd than a comprehensive list of all diseases to which the sheep is subject, so I have confined myself to mentioning the main ones, and, far more importantly, outlining how they are prevented.

Foot-and-Mouth

Two diseases are notifiable. Foot-and-mouth, with symptoms of high temperature (normal is 104°-105°F/40-40·6°C), salivating and lesions between the claws causing lameness, is distressingly sensational. On the slightest suspicion you *must* call your vet immediately. An outbreak is highly unlikely, however, unless there is already an epidemic.

The other important notifiable disease is sheep scab. This is displayed by prolonged itching, but so are certain other conditions. If sheep, both home-bred and brought-in, are properly dipped at the appropriate times, there will be no scab.

Footrot

This is a vexing condition. The cause is an organism living in the soil. Any case of lameness is suspect, the infection showing between the claws or under the hoof. Routine walks through a footbath containing copper sulphate solution or formalin is the best control, while individual cases may be treated by aerosol spray or paste. Overgrown hoofs are pared to shape with a sharp knife or cutting pliers. In theory a footrot-free flock is a possibility, but many practical farmers disagree. Injections are now available.

Abortion

Abortion in the flock is frightening. It can happen to any individual ewe through fright or accident, including rough handling, and every year a few ewes do not carry the full term, and provide pre-lambing worries. If a number of ewes abort, veterinary advice must be sought immediately. There are a number of types of infectious abortion, and the entire year's programme is upset by an abortion 'storm'.

Bacterial diseases transmitted via the soil include *lamb dysentery, pulpy kidney disease, entero-toxaemia, black disease, braxy* and *tetanus*. All are reliably controlled by the appropriate vaccines, and combined vaccines are made.

Navel Troubles and Swayback

One worthwhile help against *navel troubles* is to dip the navel of the newly born lamb right up to the body in a small jar of iodine. *Swayback* is a nervous disorder of young lambs causing unco-ordinated movement, so that the lamb tumbles about and cannot walk properly. There is no cure. Prevention is brought about through feeding a properly balanced mineral additive to the in-lamb ewe's concentrate ration, or giving access to mineral licks. The trouble with the latter procedure is that you can never be sure that *all* sheep are taking them. Occasionally an animal dislikes the taste.

Scrapie

This is a disease of the nervous system with hereditary connections. It has been known for hundreds of years, and is invariably fatal. The sheep constantly rubs itself, and severe skin irritation results. A major complication is that symptoms seldom occur before eighteen months old, but by buying older sheep one may be reasonably confident in remaining clear. An experienced buyer knows which breeds and stocks are unlikely to transmit the disease.

Pine

A debilitating disease caused through lack of cobalt. It is prevented by feeding the appropriate block or powder.

Blow Fly Strike

Addition of dieldrin to dips virtually banished this problem, but dieldrin's very persistency led to its banning. Use of a 'fly dip' two or three times during the summer should prevent occurrence. The fly lays its eggs on the fleece, usually on dirty parts, and is most active in hot, thundery weather. The small keeper can control attacks by observation; attacks are more common among large hill flocks difficult to gather.

The blow fly's eggs hatch into maggots which literally eat the sheep alive if allowed to do so. A ragged patch of wool is found, usually round the tail, and the soiled wool must be cut away and the wound dressed with a mild antiseptic or proprietary oil, *not* with dip, which is far too strong for an open wound.

Sheep Head Fly

This is a real pest, particularly in Northumberland and the Borders. It buzzes around the sheep's head during the hours of daylight, giving it no rest. Again, the small-scale keeper can doctor his animals individually. Stockholm tar round the base of the horn on horned sheep is useful, and specialized dressings may be tried.

Liver Fluke

Associated with wet pastures or patches. The host snail requires these conditions. At one stage of its life cycle the fluke migrates through the liver and settles in the bile ducts. A combined fluke and worm drench is available. Be sure to follow *precisely* the instructions on the bottles of all drenches and vaccines. They are prepared by scientists who really know their job.

Injecting and Dosing

Regarding the actual mechanics of injecting and dosing, the simplest methods are the best. A shepherd with a thousand ewes facing him needs a time-and-motion outlook and an effective set of handling pens. The smaller sheep keeper with a score of ewes is less concerned with how long the job takes; the main concern is that every sheep receives its due allowance.

A pen just big enough for the small number concerned is ideal. Then as each is injected or drenched, it is given a temporary coloured mark from a marking stick, and put into the next pen if preferred. The disadvantage of disposing of sheep as they are marked is that the last few have more room and are difficult to catch. But have a regular system; it is exasperating to think that you remember having dosed a sheep with a certain name or markings, and then be unsure about it. Only a small dot is needed; multi-coloured daubs all over the animal are unsightly.

The loose skin behind the shoulder is the best place for injection. Push the needle firmly through the skin but no further. Select a clean place, and do the job when sheep and weather are dry if possible. Follow instructions on the bottle regarding dose. A snag for the small-scale keeper is that it may be necessary to buy a full bottle of vaccine, yet have only enough sheep to require a quarter of it. Once opened, it may not be advisable to keep the vaccine for too long. There is a strong case here for a number of buyers to band together and take their purchases in bulk.

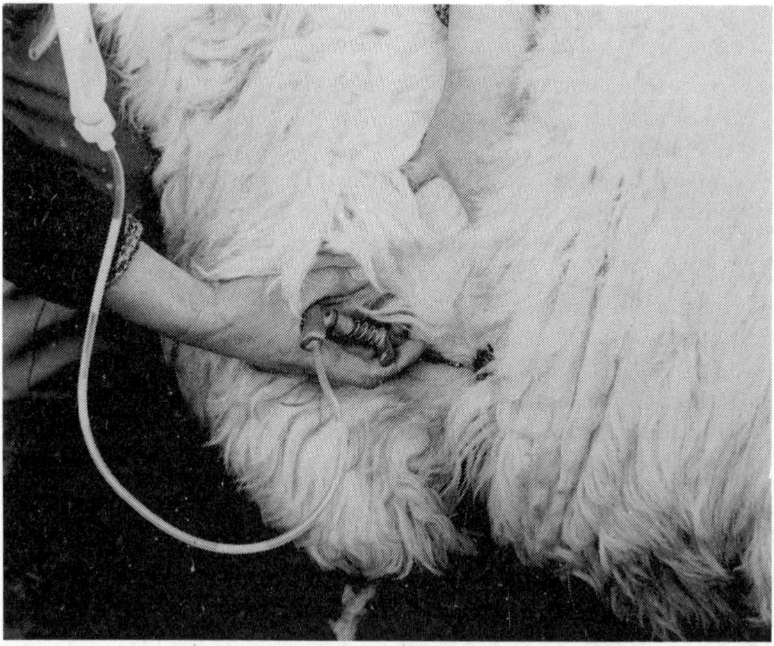

The loose skin behind the shoulder is the best place for injection.

Seldom is elaborate or expensive equipment recommended in this book, but one saving cannot be justified. Do not try to inject the whole flock using one needle. From a disease control viewpoint, there is much to be said for injecting only a few sheep per needle, and sterilizing equipment regularly. Some practical and hard-headed farmers even use a fresh needle per sheep, so that no disease can be transferred that way. A single small syringe is quite adequate, there being no point in falling to salesmen's blandishments about the benefits of large containers where only a few sheep are concerned.

All sorts of gadgets are available for drenching. None is as foolproof as the small bottle that takes one drench per sheep. You can easily find the necessary level on the small bottle, and if you wish to be methodical, acquire as many bottles as the pen holds sheep, fill them in turn from the main supply, and line them out within easy reach but well above sheep jumping height. Such a method is preferable to a so-called automatic nozzle which may work perfectly the first time, but which tends to fill with bubbles later so that you never feel quite sure that each sheep has received its due.

When dosing, be gentle. The bottle head is placed in the corner of the mouth, and tipped, slight massage on the throat ensuring that the animal swallows. Careless dosing can result in fluid down the windpipe and into the lungs, with possibly serious results. Stand by the animal's side, keep its head level, and never ram the nozzle or bottle in hard.

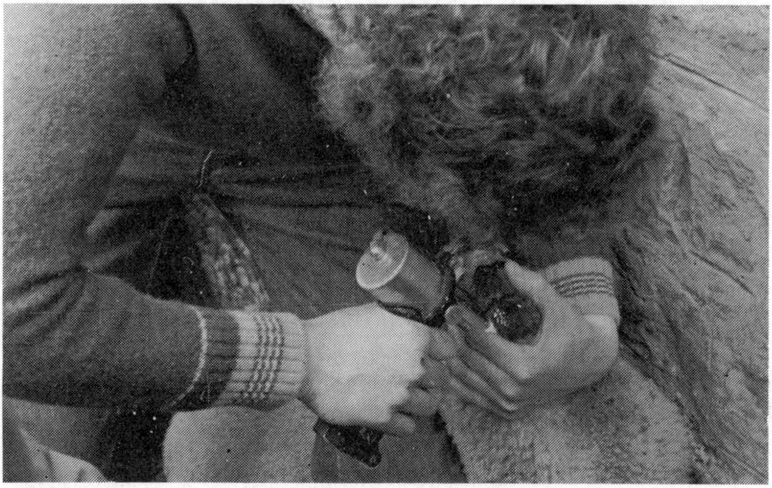

Dosing a sheep. Use the corner of the mouth, whether with bottle or gun.

10

WOOL AND SHEARING

'Wean early and shear late' is a well-proven sheep maxim. Lambs tend to compete with their dams for available pasture sooner than is generally realized, and when this happens it is better to separate the two classes and allow priority to the youngsters. But there is no point in trying to rush into shearing before nature intends.

The 'Rise'

In June or July, depending on altitude, and air temperature, the fleece tends to spring away from the skin, leaving half an inch or so of wool near the skin that is distinct from the mass of fleece. Shears, either powered or hand-operated, slice through this far more easily than if an attempt is made to take off the fleece before it is ready, when a lot of yellow, greasy matter is around. Shearing then is really hard work.

The 'rise' often accompanies hot weather. If you can choose your time, shear in the early morning or evening. The job is made far more arduous by the heat of the sun. Sheep should be shorn on comparatively empty stomachs. It is better for them, and better for the shearer. Their wool must also be dry. Therefore house them overnight without food if shearing next morning, or early in the morning if shearing during the evening. Fleeces *must* be quite dry when the sheep go inside; they need fresh breezes to remove moisture.

A slatted floor is ideal for this overnight stop, but not essential. On no account must the flock be brought on to straw, hay or chaff bedding, or the fleeces next morning will be impregnated with foreign matter tedious or impossible to remove.

Green material is far better. Stinging nettles are surprisingly effective, and neither they nor grasses stick to the fleece as dry stuff does. As the sheep keeps its own fleece perfectly clean throughout the year, if given half a chance, it is criminal to spoil the twelve month's growth through carelessness at shearing. Yet it happens time and again, and not just among beginners.

For the actual shearing, a wooden platform big enough to work on in comfort is ideal. One County Durham family of shepherds cuts squares of clean turf and makes a shearing area with them. The boards must start really clean, and be kept clean. If wool is meant for sale, the merchant supplies a sheet, which is suspended from beams and packed as shearing proceeds. Owners of more than four sheep must sell all fleeces through the British Wool Marketing Board, and buy back their requirements.

Hand shearing in the Lake District. The shearing stool saves backache, and is especially valuable where little sheep, such as the Herdwicks shown here, or Welsh, are concerned.

For home use, each fleece is best put on a table after shearing, and then packed away in a dry store – an attic suffices.

Power shears are not worthwhile for under a score or two of sheep, and to call in a contractor defeats the aims of self-sufficiency. Hundreds of thousands of sheep have been clipped with hand shears, and a good pair kept in proper order will serve well. It may pay to have them professionally ground until you have time to master the art; to clip a struggling sheep on a boiling hot day with a bad pair of shears is likely to put anyone off shepherding for ever!

Not that there need be a struggle. For more than half a dozen sheep it pays to have a catching pen near the shearing boards, complete with spring gate. The door opens inwards and springs shut on the catcher, allowing him both hands free. A horned sheep is caught by the horn, close to the head to lessen the danger of the horn breaking on an older sheep. A polled animal is 'jockeyed' out by straddling it, or drawn out by a hind leg. A catching pen should be small, just big enough to hold six or ten sheep, or any convenient number. Though these refinements are not essential at the start, they do free time and energy for other work, and so are a target for eventual achievement.

Having dragged (or been dragged by) the sheep onto the board, the next step is to sit it on its rump. This is done by placing a hand under the neck and another on the lisk or top of thigh, and either lifting (if small) or rolling (if large) the sheep into an upright position. Push the head down to the lefthand side, for a righthanded shearer. The animal will sit surprisingly quietly in this correct position.

Shear the offside of the face. Then start on the brisket, where the wool is padded hard, and move to the belly, going across and across and taking care not to pierce the vulnerable skin with the point of the shears. Take great care not to nick the pizzle of a ram or wether or the teats of ewes. Lean the sheep over slightly onto its offside, take the tail wool if there is one, and clip the part around the tail.

After shearing inside the near hind leg, that leg is straightened by thrusting the clenched fist into the flank. This, to the uninitiated, apparently miraculous posture saves time and, more importantly, reduces risk of nicking the skin. Tight skin is the aim, whatever part of the sheep is being shorn. Take off the wool well up the hind leg, and clip the tail wool if that has not been done.

Sit the sheep up, leaning it against the knees, and shear the off fore leg, neck and head on that side. We are now ready for the 'long blows', achieved by lying down the sheep on its angled back and shearing the full length of the body, from shorn hind leg to shorn neck.

Continue right over the spine; the reason for this becomes clear when finishing off.

Only one side remains to be shorn. Start at the head and neck, pushing in the knee to keep the sheep's body bowed (remember the taut skin). Take the off fore leg, (easily nicked) then around the ribs, continuing down till the hind leg and tail are reached, curving the body further and further till the task is completed.

Though there are different styles of shearing, the method described allows control over the sheep. There are few things so exasperating as to half-shear a nice clean fleece and then have the animal scramble wildly to its feet, tearing around through straw and rubbish and trailing its fleece behind like a bridal train. By far the best way to achieve proficiency is through attending classes under qualified instructors, but till such a chance occurs the beginner must try to learn step by step.

If your sheep have identifying fleece marks, these must be replaced in the precise position of the old marks, and naturally in the same colour. Some shepherds like to have a good look at feet and condition during shearing, but that is more applicable to hill farms where the sheep are only handled at infrequent intervals.

Rolling

After shearing comes rolling. The fleece is picked up with a hand on either side, where the fleece covered the hind legs, and flung, smooth or skin side downwards, onto a clean table or other clean surface. The act of flinging is done in order to rid the fleece of dust, locks and foreign matter. It is then laid out flat and even.

Flip the length of each ragged side towards the middle, leaving a neat rectangle some eighteen inches wide. Starting at the tail, roll this rectangle tightly into a Swiss roll type of bundle, till only the neck wool is left. Placing a knee on the rolled fleece, twist the neck wool into a band long enough to encircle the fleece. Wrap it round, then tuck the end under. Properly done, this wrapped fleece will withstand the several throwings and handlings encountered on its way through grader and buyer to the mill. If needed at home, it will be a pleasure to pack and store in small space. A slackly rolled fleece is an abomination.

If through any mischance the wool has become wet, it *must* be hung up to dry before wrapping. Otherwise it will deteriorate and become useless for spinning.

'Daggins' or soiled wool from around the tail should be packed separately. They should be sold or hung in a tub to make liquid manure. Never try to store any wool in polythene sacks; it will rot if it cannot breathe.

Self-sufficiency really means something when it involves growing a fleece on one's own sheep on one's own land, shearing it, spinning the wool and then knitting or weaving it into a wearable garment.

The Bradford Wool Count

Understanding the Bradford wool count helps in selecting sheep for certain purposes. Wool trade terminology is very complicated, the count being the sorter's estimate of the number of hanks each containing 560 yards of yarn that can be spun from 1 lb (450g) of combed wool. Obviously the finer the wool, the higher the count, and that basically is all that need concern the wool grower. Though the numbers range from 28's to over 80's, there is little wool under 40's, and only very fine ones like Merinos over 60's. Southdown and Hampshire Down are 56's-58's.

Each fleece is formed of basic staples or locks, the longest fibre forming the point of an apex, the shorter and usually finer fibres making the bulky base. A longwool fleece has long, narrow staples which hang in ringlets. Staple lengths vary between breeds from 1 inch (2cm) to 16 inches (40cm), but length of wool contributes only about 20 per cent of its 'spinnability'.

The crimp or waviness of wool is one of its unique features. The finer the wool, the more crimps per inch, and, because of this, sorters of Merino wool use the number of crimps per inch to assess its quality.

Wool keeps the body warm in cold weather, and cool in hot weather. It is a mistake to think that sheep suffer if they remain unshorn in a heat wave; the reverse occurs.

For hand-spun yarns used in knitting and crochet work, 54's to 58's quality has been shown to be best. The ideal length is 3 to 4 inches (8-10cm), but could range from $1\frac{1}{2}$ to 6 inches (63-150mm) and more. Crimp, softness of handling and freedom from kemp are other desirable features. Shortwool and medium wool breeds fulfil these requirements, Ryeland, Clun, Portland and Shetland all being suitable.

Loghtan, St Kilda and Jacob wools spin well direct from the fleece, while Shetland needs combing. Both Soay and Orkney wool needs so much preparation that they are difficult to spin. (v. Sylvia Hayes, *The Ark.*)

In the same survey woven fabrics – rugs, tweeds and finer cloths – are listed. The breeds already mentioned are suitable, plus Border Leicester, Derbyshire Gritstone, Romney and Whitefaced Woodland. Counts required are 48's to 56's, while minimum length is 4 to 5 inches (10-12cm). Finer fleeces from Scottish Blackface, Herdwick and Swaledale may also be used in tweeds, and coarser fleeces in rugs, according to Ryder. He adds that crimpy Down wools spin well, while lustre longwools tend to be 'flat' and lacking in resilience, making them difficult to spin. They may, however, be used in tapestry weaving.

Spinning

When spinning with a hand spindle, a length of fibres is drawn out, twisted by hand and attached to the top in a notch, hook or half-hitch. After some spinning, the length of yarn spun causes the spindle to reach the ground, when the spun yarn is wound onto the spindle and the process repeated. A hand spindle is easily made from a 3/8 inch (9mm) rod, or even from a green twig. A notch is cut near the upper end to hold the yarn.

Dyeing

There are limitless opportunities for the 'home grown' enthusiast to brighten his wool or cloth. The range of natural dyes is vast, and varies according to locality. Dyes are of two main kinds. The substantive ones may be applied direct to the fibre from a solution, whereas the adjective section requires a mordant or additional chemical to attach them to the fibre.

Alum is a common mordant, and is used with many plant dyes. Mordants combine with both wool and dye, making the colour completely fast. It is essential to scour wool thoroughly before applying any dye, as dirt or grease prevents the dye penetrating the fibre.

Mordants are used in the proportion of one to five, i.e., 3 to 4 oz. (75-100g) per lb (450g) of wool. A smaller amount, say 1 oz. (25g), of cream of tartar is used with the alum to brighten the final shade, and make it more even.

For each pound of wool, 1 to 2 lb (450-900g) of plant is required. Dye is yielded from almost any part of a plant, including fruit, leaves, flowers, bark, stem and root. The part used by the dyer varies with different species.

Onion skins are a well-known source, being easy to apply and leaving an attractive golden-yellow. Blackberries give blue dye, heather flowers an olive brown, while a different shade of brown comes from ash bark. For green, broom leaves are suitable, while ragwort flowers give a yellow dye, and dandelion roots yield purple. Lichens gave the original rich brown dye of Harris tweed. The lichens are most effective if soaked in dilute alkali (ammonia) beforehand. Stale urine once provided the alkali. Then boil lichens and wool together, the former in a cotton bag.

A metal bucket is used for mordanting and dyeing. The alum and cream of tartar are added to two to three gallons of cold soft water, and dissolved while the water is heating. When the water is lukewarm, 113°F (45°C), add the wool and bring the whole to the boil gradually. Simmer for thirty to forty-five minutes. Lift out the wool and squeeze.

Dye is extracted from plants by adding cold soft water in a metal bucket, bringing to the boil and boiling until the colour has left the plants. This may take only a minute or two for fruit, or several hours for bark. Strain the liquid to remove the plant material, then cool.

For the dyeing process, the wet wool is added to the cold dye solution, which is brought slowly to the boil. The mixture is stirred frequently for about an hour while it simmers, but this period is only a rough guide. Simmering and stirring continue till the required shade is attained. Wet wool looks darker than dry, so a little wool is removed on a stick and squeezed dry to assess the true colour. Wash the wool in running water and spread out to dry.

The 'finishing' process involves treading the wool in a tub of water, using bare feet, and cleansing is done by adding fuller's earth.

Preparation of animal skins to make them soft, pliable and long-lasting is one of man's oldest arts. The length of wool on the skin is determined by shearing date, or in the case of lambskin, no shearing is done. The butcher may be asked to salt the hides or put them in a refrigerator, after which they are sent to a tannery. Though this is better than not using the skins, most self-sufficiency pioneers wish to preserve their own.

Tanned sheepskins vary in colour according to the kind of wool on the sheep. They may be white, grey, brown or mottled. Sheepskin jackets are best made from lambskins, moccasins and car seat covers from the heavier hides. Rugs and pillows are another possibility from mature sheep. Special needles are used in make-up, and a suitably strong thread is essential.

Tanning

One of the early methods was to soak the skins in a tannin solution. Bark-tanned skins are brownish in colour, chestnut giving a dark brown and oak a lighter shade.

Alum tanning is recommended for home use. It is safe, simple, and does not require large capital investment. Skins from butchered animals can be preserved until enough are on hand by salting each liberally and placing on wooden slats to dry. One pound (450g) of salt per skin is needed.

A large tub or stone crock is needed for the actual tanning. Soak each skin for two hours in lukewarm water to soften it, then rinse until the water is clean. The skin is then washed in laundry soap or detergent. Fat or flesh is scraped off during this process, and as much water as possible is removed by squeezing in towels.

Alum and salt are now used. The hide, fleece side down, is rubbed with a mixture of 4 oz. (100g) alum and 8 oz. (225g) salt, or soaked in a solution of 4 oz. (100g) alum and 2 oz. (50g) salt per gallon ($4\frac{1}{2}$ litres) of water. Turn the hide twice a day for three days, ensuring that it is completely covered in solution throughout.

If no solution is used, rub in the alum and salt and leave overnight with the mixture on. Next morning, hang over a rail skin-side out. On the following day the mixture is sponged off, and the skin softened by rubbing in a little neat's-foot oil or glycerin. The skin must be worked as it dries, otherwise it will become hard. Draw it backwards and forwards over a table edge.

The pelt should be sanded after it has dried and softened. Egg cleaning or electric sanders may be used.

All spare bits of lambskin or sheepskin should be saved. They are useful for car cleaning (outside), polishing woodwork, lining boots and shoes, as heavy shoulder straps and for bicycle seats. Some pieces will be too small even for these purposes, but if put into a suitable box they make an ideal cat bed.

Preventing Skin Faults

A sheepskin is a valuable commodity. It has a variety of uses, including rugs, but its quality is variable. Some skin faults are obscure, but most are preventable by the farmer at no cost.

Injection abscesses form as a result of infection after injecting the sheep. After healing, the structure of the skin is changed and the part dyes differently or produces a different suede nap. Scar tissues are of similar origin and the effect is equally preventable. All that need be done is to inject in the loose skin behind the shoulder. Then if an abscess does form, it is on the edge of the pelt and is easily trimmed off, whereas if in the centre of the back, the pelt is greatly reduced in value.

Other precautions are to inject in dry weather on dry sheep. Sterilize needles as often as possible, and always if any contamination is known to have occurred. Injection of young lambs can be reduced to a minimum by adopting a full policy of vaccination to the ewes in late pregnancy.

11
THE SHEEP DOG

Where a few sheep are kept in a paddock it is by no means essential to own a sheep dog. Your stock become so used to your company that they will come at a call, or be easily enticed into a holding pen. At the other extreme, do not imagine that half a dozen sheep are enough for dog training, and that a succession of highly-schooled Collies will give additional momentum to the holding. Sheep that are continually herded by dogs on the same small area of ground become so accustomed to what is required that they make either for the pen or the handler as soon as the dog appears. It is possible to train dogs when owning only a few sheep, but that is done by constantly selling and reselling. It has little part in a self-sufficiency programme.

For people with a few fields a sheep dog is a tremendous help. For those with hill grazing it is almost an essential, unless you wish a neighbour to gather your sheep for you, which is self insufficiency. The BBC2 programme *One Man and His Dog* has aroused enthusiasm among many who previously believed that working a sheep dog was similar to the occult.

The breed in Britain is the Border Collie. You may buy a pup, and do all the work on it yourself. While no pup can be *guaranteed* to work sheep, the chances of success are far greater if one is bought from parents both of whom work sheep. The pup need not be of registered stock, but dog owning has wider implications if it is.

If you wish to breed, the pups will be worth more if registered, or be more appreciated by friends. First step is to join the International Sheep Dog Society, whose secretary is Mr Philip Hendry, 64 St Loyes Street, Bedford. Until you join ISDS you cannot register pups.

Except in rare circumstances, pups may be registered only if both parents are registered. Within a fortnight of the mating, the sire's owner should send a mating card to ISDS, and the dam's owner will then receive a registration folder. When pups change hands, markings must be recorded and new owners' names notified to ISDS within six months of the pup's birth. The Society will inform you of names and addresses of breeders in your area. Beware like the plague any so-called Border Collies registered with the Kennel Club only.

Helping a weak or small new-born lamb to suck. The dog arouses the ewe's maternal instincts. Sheep is a Scottish Blackface.

Puppies should be wormed at three, five and seven weeks of age according to present veterinary advice, and then every two weeks until six months old. After that, twice a year against round worms sufficies; dose against tape worm only when the flat, whitish segments are seen in the dung. It is to everyone's interests to keep dogs free from external parasites – fleas and lice.

The first innoculation against distemper, hard pad and leptospirosis is given by the vet at twelve weeks, the second at fourteen weeks. Keep away from other dogs until a fortnight after the second injection; thereafter give a booster once a year.

So much for the preliminaries. The best feed is a proprietary meal twice a day from weaning to six to ten months according to growth, once daily thereafter.

A Border Collie's instinct for work should show itself from a few months' of age up to eighteen months or two years. The timing is no criterion of eventual prowess. It will begin to round up poultry, cats or other dogs, and sheep if allowed. Training should not start until the trainee has strength enough to outpace the fastest sheep; if used too young, it will merely become exasperated through being unable to keep up, and develop bad habits such as barking.

Training

The pup may be accustomed to a collar at an early age. It should then be taught to lead, using something that it cannot bite through. By now the dog will have a name. Use it. Call the youngster to you and praise it *as soon* as it responds and turns towards you, not merely when it arrives. A well-bred Border Collie is a highly sensitive machine, with hearing far superior to human's. Bawling and shouting at a dog is unnecessary and a reflection on the handler, not on the dog.

Next lesson is to Lie Down and Stay Down. It should not be taught too young, or the pup's inclination to return to its handler may be retarded, and its individuality set at risk. Give the command and hold it firmly down by the bottom of the collar. At first the dog will want to follow your movement as you stand up. Firmness and patience are required – the dog must learn to stay down if you command it to do so. *Always* give the release command, usually a quick double pat on the shoulder. All exercises *must* have a finish, so the dog knows when it is off duty.

A few quiet sheep in the corner of a small paddock or yard are used for early training. A large field of wild sheep is useless. Walk the sheep into one corner, release your dog and let it run towards them. Anything may happen, or nothing. In its excitement the youngster may even forget its 'Stop' command, and bustle the sheep roughly.

Ideally it will move from one side to the other of the cornered flock, preventing any from escaping. A well-shepherded flock is a great help in training a young dog, as the ewes respond to each movement. As the dog moves to *its* left, give the voice command for that movement, often 'Away To Me'. 'Come Bye' is commonly used for the dog moving to *its* right, but the actual words are of no significance. You may borrow some, or make up your own, provided you are consistent.

In time the dog associates its movement with your command, until the day comes when it moves in a certain direction *because* of what you have said. This training may take weeks; no timetable can possibly be attempted.

The other necessary command is 'Walk Up'. The dog should walk straight forward towards its sheep. These five basic commands should cover all everyday jobs among sheep. The commands are:

Lie Down (or *Stop*)
Return to handler's side (by calling dog's name)
Go Left
Go Right
Walk Up

A well trained Border Collie is a great friend and an abiding source of satisfaction. Nothing will help a newcomer fit into a farming neighbourhood better than willingness to go along and help with a dog that knows its stuff.

GLOSSARY

GENERAL SHEEP TERMS

Allotment Walled enclosure, usually on a hillside, and of better quality than the hill above.

Barren, eild, geld A ewe not in lamb.

Broken-mouthed Older sheep with loose or missing teeth, unable to graze properly and therefore requiring easily-grazed herbage.

Cast An old sheep fit only for the butcher. In some districts, a sheep fast on its back and unable to rise.

Couped, rig-welted A sheep turned onto its back and helpless

Crone Old ewe.

Crop Preceded by a number, denotes the number of lamb crops that the ewe has had.

Draft ewe At four to six years old, the hill ewe is drafted downhill for further breeding.

Ewe Adult female.

Fell. hill, moor Regional name for upland areas of natural grazing, sometimes unfenced.

Gimmer Young female. A gimmer lamb is a ewe lamb.

Heft An area of hill grazed by certain sheep, who become 'herded' to that part.

Hirsel A number of hefts, comprising the area of ground herded by one shepherd, or the sheep on them.

Hogg, teg (also *hogget, hoggerel*) From December of the sheep's first year to its first shearing the following summer.

Inbye The fenced or enclosed lower fields of a hill farm.

Lamb Young sheep, from birth to weaning or the end of its year of birth.

Lunky hole, smout hole Specially made low hole in a stone wall; takes only one sheep at a time, and is easily blocked when not in use.
Polled Hornless.
Ram, tup Interchangeable terms for an uncastrated male.
Rud, raddle Coloured powder, the base of a marking paste.
Spain, wean The act of taking young stock from their dams. End of the milk-feeding period.
Strike, struck Shepherds' terms for attacks of the blow fly, maggots from whose eggs attack the living flesh.
Theave As gimmer.
Twinter Sheep in their second winter.
Two-shear Sheep in its third year. Most are first shorn at just over one year old.
Two-tooth As gimmer.
Wether Castrated male for eventual fattening.

WOOL TERMS

(from Michael Ryder's *Sheep and Wool for Handicraft Workers*)

Break Tender part of staple.
Britch or breech Coarse wool from hind legs.
Cast Fleece that does not reach grade standard.
Character Regularity and definition of crimp.
Count Degree of fineness (see *Quality*).
Crimp Natural waviness of wool fibres.
Crossbred Wool less than 60's quality, not necessarily from crossbred sheep.
Daggings, doddings Wool from around the tail area, containing solid lumps of dung.
Dead wool Misleading term to describe hairy fibres; all wool is dead!
Deep Term applied to long wool.
Density Closeness of fibres to one another in the staple.
Grade Standard on which wool is bought and sold.
Grease The skin secretion lanolin.
Handle Degree of softness.

Kemp Short, very coarse fibres. Appear chalky white and have pointed tip and tapering root.

Quality, count, number Degree of fineness.

Rise Spurt of wool growth in spring.

Staple Natural lock or tuft of fibres.

Strong Coarse wool of low quality.

Suint Deposit on fleece from sweat gland.

Woolen Yarn made from carded wool.

Worsted Yarn made from wool that is combed after carding to make the fibres parallel.

Yolk Yellow discoloration of fleece caused by mixture of grease and suint.

FURTHER READING

The Ark (monthly). Rare Breeds Survival Trust, Winkleigh, Devon.

Bowen, Godfrey. *Wool Away!* Whitcombe and Tombs, 1955.

British Sheep. National Sheep Association, 1976.

Cooper and Thomas. *Profitable Sheep Farming*, Farming Press, 1972.

Firbank, T. *I Bought a Mountain*. Harrap, 1940.

Fraser, Allan. *Sheep Farming*. Crosby Lockwood, 1937.

Grant, D., and Hart, E. *Shepherds' Crooks and Walking Sticks*. Dalesman, 1972.

Hart, Edward. *Northcountry Farm Animals*. Dalesman, 1976.

———. *Showing Livestock*. David and Charles, 1979.

———. *The Hill Shepherd*. David and Charles, 1977.

Longton, T., and Hart, E. *The Sheep Dog: Its Work and Training*. David and Charles, 1976.

Morrison, A.M. *Red Dragon Farm*. Faber, 1964.

Nixon, David. *Walk Soft in the Fold*. Chatto and Windus, 1977.

Perry, Richard. *I Went A-Shepherding*. Lindsay Drummond, 1944.

Rainsford-Hannay, Col. F. *Dry Stone Walling*. Faber, 1957.

Raistrick, Arthur. *The Pennine Walls*. Dalesman, 1972.

Rollinson, William. *Lakeland Walls*. Dalesman, 1972.

Ryder, Michael. *Sheep and Wool for Handicraft Workers*. (Available from 23 Swanston Place, Edinburgh.)

USEFUL ADDRESSES

Animal Breeding Research Organisation, West Mains Road, Edinburgh.

British Wool Marketing Board, Oak Mills, Clayton, Bradford, W. Yorks.

Country Landowners' Association, 16 Belgrave Square, London SW1.

Grassland Research Institute, Hurley, Maidenhead, Berkshire.

Hill-farming Research Organisation, Bush Estate, Penicuik, Midlothian.

Institute of Animal Physiology, Babraham, Cambridge.

International Sheep Dog Society, Bedford.

International Wool Secretariat, Wool House, Carlton Gardens, London SW1.

Joint Sheep Group, National Farmers' Union, Agriculture House, Knightsbridge, London SW1.

Meat and Livestock Commission, Queensway House, Bletchley, Milton Keynes.

National Institute of Agricultural Botany, Huntingdon Road, Cambridge.

National Sheep Association, Jenkins Lane, St Leonards, Tring, Hertfordshire.

Breeders' Associations

Blackface, 24 Beresford Terrace, Ayr.

Black Welsh Mountain, Summer Lane, Combe Down, Bath, Avon.

Blue-faced Leicester, Blencogo, Wigton, Cumbria.

Border Leicester, 24 Beresford Terrace, Ayr.

Cheviot, 5 Tower Knowe, Hawick, Roxburghshire.

Clun Forest, 11 Blackfriars Street, Hereford.
Colbred, Crickle Barrow, North Leach, Cheltenham, Gloucestershire.
Dalesbred, 45 Allhallows Lane, Kendal, Cumbria.
Dartmoor, Bilberry Hill, Buckfastleigh, Devon.
Derbyshire Gritstone, Red Lees Road, Cliviger, Burnley, Lancashire.
Eppynt Hill and Beulah Speckled-face, Market Street, Builth Wells, Powys.
Exmoor Horn, The Avenue, Minehead, Somerset.
Herdwick, Glenholm, Penrith Road, Keswick, Cumbria.
Hill Radnor, Newmarket Chambers, Abergavenny, Gwent.
Jacob, Jenkins Lane, St Leonards, Tring, Hertfordshire.
Kerry Hill, Milford Road, Newtown, Powys.
Llanwenog, Pertheos, Creuddyn Bridge, Lampeter, Dyfed.
Lonk, Jack Hey Lane Farm, Cliviger, Burnley, Lancashire.
North Country Cheviot, Bellfield Road, Kessock, Inverness.
Rough Fell, Milnthorpe Road, Kendal, Cumbria.
Shetland, Fairview, Vidlin, Shetland.
South Wales Mountain, The Bryn, Pontilanfraith, Gwent.
Swaledale, Spring End, Low Row, Richmond, North Yorkshire.
Teeswater, Edengate, Warcop, Appleby, Cumbria.
Welsh Mountain – hill, c/o WAOS, Brynawel, P.O. Box 8, Aberystwyth, Dyfed.
Welsh Mountain – pedigree, Tudor Cottage, Llwyndafydd, Llandysul, Dyfed.
Wensleydale Longwool, 27 Fountain Street, Ulverston, Cumbria.
White-faced Dartmoor, Sawdye and Harris, Newton Abbot, Devon.
Welsh Halfbred, Brynawel, Aberystwyth, Dyfed.

Manx Loghtan North Ronaldsay or Orkney Portland St Kilda Soay White-faced Woodland	Combined flock book, Lower Eastrip Farm, Colerne, Chippenham, Wiltshire.

INDEX